Brutes or Angels

Brutes OR
Angels

HUMAN POSSIBILITY
IN THE AGE OF
BIOTECHNOLOGY

JAMES T. BRADLEY

THE UNIVERSITY OF ALABAMA PRESS
Tuscaloosa

Typeface: Perpetua and Helvetica

Cover art: *man x-ray,* ©iStockphoto.com/cosmin4000;
Genetic Rain, ©iStockphoto.com/mstay.
Cover design: Erin Bradley Dangar / Dangar Design

∞

The paper on which this book is printed meets the minimum requirements of American
National Standard for Information Sciences—Permanence of Paper for Printed Library
Materials, ANSI Z39.48-1984.

Library of Congress Cataloging-in-Publication Data

Bradley, James T., 1948–
Brutes or angels : human possibility in the age of biotechnology / James T. Bradley.
pages cm
Includes bibliographical references and index.
ISBN 978-0-8173-1788-1 (trade cloth : alk. paper) —
ISBN (invalid) 978-0-8173-8658-0 (ebook)
1. Genetic engineering—Moral and ethical aspects. 2. Human Genome Project—
Moral and ethical aspects. I. Title.
QH442.B72 2013
174.2—dc23
2012035096

For my grandchildren and their descendents

Contents

Illustrations

TABLES

Preface

It should be the aim of every scientist to eventually generalize his [her] views of nature so that they make a contribution to the philosophy of science.

—Ernst Mayr, *This Is Biology: The Science of the Living World*

The objective of this book is to facilitate informed choice making about personal use of biotechnologies and formulation of public policies governing their development and applications. The book provides basic information about a wide range of biotechnologies, the ethical issues raised by each one, and diverse viewpoints on dealing with these issues. Two underlying premises imbue the book:

1. Earth's life forms are products of nature and share a common ancestry extending back about 3.8 billion years.
2. Life's processes are coming increasingly under human control. Appreciating how life got to where it is today can foster wisdom in choosing its future directions.

No formal learning in biology is needed to read and understand this book. In chapter 1 the basic principles of cell and molecular biology are laid out for readers wishing to begin with scientific fundamentals underlying the biotechnologies described in succeeding chapters. Chapter 1 tells how molecules and cells are organized into living things and why it matters to know something about how cells work. Chapter 1 aids and enriches understanding of later material, but some readers may choose to begin with a later chapter, exploring straightaway a specific biotechnology with its ethical, legal, and societal implications: stem cell research and therapy, embryo selection, the Human Genome Project and human genomics, genetic enhancement, cloning, age retardation, cognitive enhancements, or synthetic biology. Chapters need not be read chronologically.

Chapters 2–9 address biotechnologies that directly involve human beings at the levels of developing embryos, cells, genes, and brain function. For example, chapter 5, on human genomics, includes special sections on genes and ethnicity that explore the benefits and risks of using genomic information about human diversity to advance health care and disease prevention. Chapter 10 examines the burgeoning field of synthetic biology and our emerging role as creators of life itself. Each chapter has two objectives:

1. to present the science required for making informed choices about developing and using a particular biotechnology and
2. to encourage personal reflection on human values as they relate to the biotechnology.

Metaphors, analogies, drawings, and photographs help convey information about important biological principles and biotechnological procedures. I strive to present multiple views on the ethical issues without bias. However, when a particular viewpoint is based on the misuse or misunderstanding of scientific information, I point that out.

The theme of the book is that the interface between modern biotechnologies and human values is very dynamic, changing as biotechnologies develop and as individual human minds and entire societies respond to their understanding of those biotechnologies. Different technologies raise different ethical issues. For example, defining personhood is the major issue facing use of human embryos for embryonic stem cell research; the meaning of personal identity arises with reproductive cloning; privacy and ethnic discrimination are issues for human genomics; the meaning of human authenticity and human nature crop up with neuroenhancement and genetic enhancement; eugenic issues arise with preimplantation genetic diagnosis; the biological and societal significance of death occurs with age retardation; and appropriate actions and responsibilities for creators of new life forms accompany synthetic biology.

All twenty-first-century biotechnologies bring up the issue of *distributive justice.* Who will benefit most from these technologies, and will the benefits be accessible to people who need them the most? How should finite public funds and research time be used? Discussion of these and related questions of distributive justice occur throughout the book. Some writers prefer to use the term *moral* when referring to religion-based codes for behavior and *ethical* for secular-based guidelines. I do not make this distinction but instead use the two terms interchangeably, as do most bioethicists.

At the end of each chapter is a set of questions for thought and discussion suited for personal reflection and group dialogues. Most of the questions have no single, correct answer; rather, the questions are important for the public discussion needed to shape wise policies at state, national, and international levels.

A comprehensive glossary follows the epilogue. Although nontechnical language is used throughout the book, certain terms like *molecule, protein, DNA, gene, blastocyst, germ cell,* and *zygote* are needed for clarity and accuracy when discussing a particular biotechnology and the ethical issues arising from it. Each such term is italicized and clearly defined when it first appears in the text. Then, when it first appears in a later chapter, it is italicized again to signal that its definition appears in the glossary. Notes and sources for additional information are included for each chapter, and a comprehensive reference list and detailed index complete the book.

Acknowledgments

Many persons contributed ideas and tangible support for this project. Sue Bradley, Isabelle Thompson, and Amy (Gordon) Jones read and edited major portions of the book before the manuscript was submitted for publication. Christine Probst, Alondra Oubre, Lana Saffert, and the late Martha King read chapters on cloning, the genomics of human diversity, neuroscience, and stem cells, respectively, and gave valuable advice for improving them. I am also grateful for numerous and regular conversations about biotechnology and bioethics with many persons over many years. These conversations helped me to see what things were most important to write about and how best to write about them. Insights and questions from my wife, Sue Bradley, were invaluable in this department. Others contributing time and illuminating conversation include, but are not limited to, Ray Allar; my mother, Marge Bradley Scholl, and late father, Donald L. Bradley; my daughters, Laura Bodt and Sara Compaglia; John Bodt; my sister, Marilynn Bradley; John Compaglia; Gerry Elfstrom; Keith Gibson; Franco Giorgi; Robert Greenleaf; Peter Harzem; Katie Jackson; Jay Lamar; Clark Lundell; Anthony Moss; Gary Mullen; Leonard Ortmann; Randy Pipes; Teresa Rodriguez; Michelle Sidler; Michel Smith; Timothy Terrell; Frank Werner; the sisters in P.E.O. International, Chapter E, Auburn, Alabama; the leaders and participants in the 2003 Intensive Bioethics Course XXIX at the Kennedy Institute of Ethics at Georgetown

University; my genethics course students at Auburn University; and Tim Turner's research ethics students at Tuskegee University. I am grateful to the science and mathematics teachers who inspired and nurtured my interest in those ways of knowing, from junior and senior high school in Rice Lake, Wisconsin (George Theis, James Stauffer, Thomas Ritzinger, Karl Schmid, and Wayne Arntson), through my education at the University of Wisconsin (notably, C. H. Sorum and James F. Crow). And I owe unique debts of gratitude to my major professor at the University of Washington, the late John Edwards, for demonstrating the value and fun of expanding one's creative endeavors beyond a single, narrow scientific niche; to molecular biologist and bioethicist Kevin Fitzgerald at Georgetown University for advising me that a cell biologist can contribute to the discipline of bioethics by simply "doing it"; and to the late evolutionary biologist, science historian, and author Ernst Mayr (1904–2005) at Harvard University for challenging and encouraging me to write for a general audience about the philosophical significance of cell biology.

I am fortunate to have Janna Sidwell as my primary illustrator. Her patience with me, professionalism, dedication to the project, and artistic talent are largely responsible for making the biological concepts accessible to non-biologists. Sara Agnew also contributed to some illustrations, and undergraduate Erin Beebe helped research many topics. I thank my department chairs, James Barbaree and Jack Feminella, and my dean, Stew Schneller, at Auburn University for encouraging my foray into bioethics and book writing. I also thank Elizabeth Motherwell, senior acquisitions editor for Natural History and the Environment at the University of Alabama Press, and freelance editor Karen Johnson and project editor Jon Berry for the many ways they supported this project—with encouragement, work, and wisdom. The constructive critiques of three anonymous reviewers were also invaluable. Finally, my deepest gratitude goes to my wife, Jackie Sue, for her constant encouragement and patience with my geographical absences and mental distraction while writing.

Brutes or
Angels

Introduction

Who is there that does not wonder at man? . . . [M]an fashions, fabri-
cates, transforms himself into the shape of all flesh, into the character
of every creature.
    —Giovanni Pico della Mirandola, "Oration on the Dignity of Man"

Humankind's place in nature is somewhere between brutes and divinities
according to Renaissance philosopher Pico della Mirandola (1463–1494).
Writing what some consider to be the manifesto of the Renaissance, Pico
allegorized humankind's creative potential. He imagined God telling humans how they differ from other creatures:

> The nature of all other beings is limited and constrained within the
> bounds of laws prescribed by Us. Thou, constrained by no limits, in
> accordance with thine own free will, in whose hand We have placed
> thee, shalt ordain for thyself the limits of thy nature . . . with free-
> dom of choice and with honor, as though the maker and molder of
> thyself, thou mayest fashion thyself in whatever shape thou shalt pre-
> fer. Thou shalt have the power to degenerate into the lower forms of
> life, which are brutish. Thou shalt have the power, out of thy soul's
> judgment, to be reborn into the higher forms, which are divine.
> (Pico della Mirandola [1486] 2005, 287)

Pico prepared his famous "Oration" as the opening statement for a debate
in Rome during the winter of 1486–1487, but since some of Pico's theses
were ruled heretical, the debate never occurred.

Pico's fifteenth-century words are remarkable in that they describe with uncanny accuracy our twenty-first-century understanding of humankind's position in nature. Modern paleoanthropology and molecular biology tell us that *Homo sapiens* evolved some two hundred thousand years ago from "brutish" bipedal ancestors who climbed, walked, and ran in Africa nearly five million years ago. And evolutionary psychologists do little better than Pico in describing how humans differ from other animals: humans philosophize and theologize about how best to live their lives, while other animals probably do not. With the biotechnologies described in this book, humankind is poised to shape its biological future very nearly as Pico mused over five centuries ago, for better or worse, according to how we exercise our judgment.

Humankind is in an age of biotechnology. Our relationship with biotechnology during the next decade or so will largely determine the future nature of our species and the rest of the living world. *Homo sapiens'* future, whether we "degenerate" or are "reborn into higher forms," will reflect the thoughtfulness we bring to decisions about how to use twenty-first-century biotechnologies.

The inspiration for this book came in an e-mail message from my lifelong friend Ray. It was 1999. Completion of the *base* sequencing of *DNA* for one of the twenty-three human *chromosomes* had just been announced by news media. "Hey, what the hell does this mean, 'sequencing a chromosome,' and why should I care?" wrote Ray. I began a reply that included a description of DNA molecules and how they carry hereditary information for both desirable traits and genetic diseases. But, after spending thirty minutes on an abbreviated answer to his very good question, I wrote, "Ray, you ask a great question, but there's more to the answer than this. So I'm going to write a book for you about *cells, genes,* DNA, and the new biotechnologies. I hope you will find it interesting." Within minutes Ray's short reply appeared on my monitor: "Thanks, Jim, I just hope you finish it before I forget my question." The book project took longer than expected, so I've periodically reminded Ray of his initial question and, in the meantime, answered a few others.

While writing, I had not only Ray in mind, but also my musician wife, nonscientist parents, sister, daughters, friends, and students. Their questions about *stem cells, cloning, genetic engineering, in vitro fertilization* (IVF), and other biotechnological terms in the news helped me identify the book's topics.

I began this introduction with Pico, from the small village of Mirandola in northern Italy, who imagined humans between animals and divinities, with the ability and authority to develop in either direction. A twentieth-century thinker just as creative and inquisitive as Pico was Jacob Bronowski (1908–1974), who began his classic book, *The Ascent of Man,* with a chapter titled "Lower than the Angels." Here is his description of our species: "Man is a singular creature. He has a set of gifts which make him unique among the animals: so that, unlike them, he is not a figure in the landscape—he is a shaper of the landscape" (Bronowski 1973, 19).

What Bronowski did not foresee in 1973 is that in addition to shaping our environment, we would soon become sculptors of human nature itself. So here we are, early in the twenty-first-century, between brutes and angels, with burgeoning abilities to enhance, engineer, manipulate, and even create life in ways unimaginable just a few decades ago. Deciding how to develop and use these powers will not be easy, but we must do it.

Living a good life in the Age of Biotechnology requires gathering reliable information, listening carefully and respectfully to others' views, formulating one's own views thoughtfully and deliberately, and maintaining open pathways for dialogue. It also requires being open to changing one's views in light of new information, either in the realm of science or in the area of ethical argumentation. As humankind moves into the Age of Biotechnology, I am hopeful for its future in proportion to the diligence with which we exercise our skills in gathering information, conversing with mutual respect, being open to change, and formulating thoughtful opinions.

I

Cells and Molecules

The Unity of Life

> Long ago it became evident that the key to every biological problem
> must finally be sought in the cell; for every living organism is, or at
> sometime has been, a cell.
>
> —Edmond B. Wilson, *The Cell in Development*
> *and Heredity (Genes, Cells, and Organisms)*

On August 20, 1979, *Newsweek* magazine sported a cover with a beautiful color cartoon of a single *cell* and its interior. This and the accompanying story, "Secrets of the Human Cell," illustrated the prominence and relevance of cell biology for the general public that was evident more than thirty years ago. Since then, discoveries in cell biology and biotechnology have given rise to the so-called new biology that increasingly influences how we are conceived, how we live, and when we die.

Cells comprise our bodies; cells, in turn, are comprised of chemical units called *molecules.* Why does knowing about cells and molecules matter? Every biotechnology discussed in this book, from *stem cells* and *cloning* to *genetic enhancement* and *age retardation,* has its foundation in cellular and molecular biology. Quite literally, cells are us. We are made of cells, some twelve trillion of them. Not only do cells comprise or manufacture all the parts of our physical bodies, but they also make us thinking, rational, even spiritual beings. Writing about the 100 trillion cell-to-cell communication sites (*synapses*) inside the human brain, neuroscientist Joseph LeDoux (2002, ix) puts it this way, "You are your synapses. . . . They are the channels of communication between brain cells, and the means by which most of what the brain does is accomplished."

As we learn more and more about cells and the chemistry of life, we gain correspondingly more opportunities to biologically alter ourselves and the rest of the living world. Reshaping life wisely requires biologically literate non-biologists to make decisions about how biotechnologies are developed and used. This is why knowing about cells and molecules matters. Moreover, learning about the beauty of life beyond what our unaided eyes can see is fun. And life should be fun. So let's get started and jump right into the world of the cell. The following questions will guide our foray into the microscopic and submicroscopic realms of biology:

1. What are cells and molecules?
2. What do cell biologists do?
3. How is cell structure related to cell function?
4. What are the relationships between *DNA,* genes, *chromosomes,* and *genomes?*
5. What is the *central dogma of biology?*
6. What is the *genetic code?*
7. How do cells reproduce?
8. When and how did the first cells originate?
9. What do cells have to do with human values?

Cells and Molecules

Just as different types of buildings are units of a city, cells are the structural and functional units of all living things on Earth. But unlike buildings, all cells come from pre-existing cells. Together, these two statements about cells constitute the "cell theory." The first, that all living things are comprised of cells, was proposed for plants in 1838 by a German botanist, Matthias Schleiden, and for animals in 1839 by Schleiden's zoologist colleague, Theodor Schwann. The second statement, that only cells beget cells, was proposed in 1855 by the German pathologist Rudolf Virchow. Direct observations and experimental data soon elevated the original propositions of Schleiden, Schwann, and Virchow to the level of theory, with a certainty comparable to that enjoyed by the heliocentric theory for the solar system

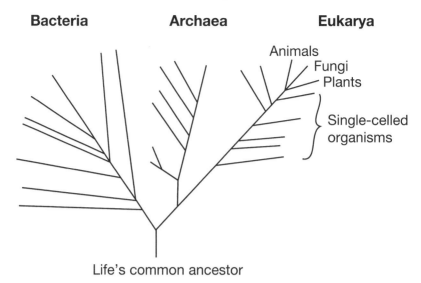

Bacteria **Archaea** **Eukarya**

Animals
Fungi
Plants

Single-celled
organisms

Life's common ancestor

Fig. 1.1. The three kingdoms of life. The prokaryotic kingdom Archaea, to which exotic single-celled organisms living in extreme environments belong, is more closely related to the kingdom Eukarya, to which humans belong, than to the other prokaryotic kingdom, Bacteria. Time moves from the bottom (about 3.8 billion years ago) upward and outward to the present in this tree-of-life diagram.

and the germ theory for disease.[1] The two components of the cell theory have now been established principles of biology for 150 years.

Broadly speaking, there are two kinds of cells: *prokaryotic* and *eukaryotic*. Prokaryotic cells include bacteria and some types of algae. In the late 1970s, prokaryotic cells were divided into two major groups based on biochemical and genetic characteristics. So now biologists recognize three kingdoms of cells: Eukarya, Bacteria, and Archaea (fig. 1.1), the latter two kingdoms being prokaryotic. Bacteria include the familiar beneficial and disease-causing microbes that live in our gut, grow in the soil, give us infections, or cause Lyme disease. Archaea live in extreme environments like the hot springs in Yellowstone National Park and in places with high salt or methane levels. Eukarya include organisms from amoebae and armadillos to mushrooms, mimosa trees, mountain lions, and humans, all comprised of eukaryotic cells.

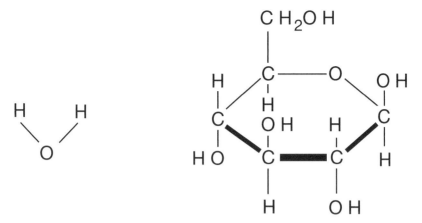

Fig. 1.2. Water (*left*) and glucose (*right*) molecules diagramed. Letters symbolize atoms and connecting lines symbolize chemical bonds.

You may wonder how viruses fit into this cell classification scheme. They do not! Viruses are not alive, and they are not cells. Viruses can cause havoc inside cells by commandeering life's normal processes and subverting them toward propagating more virus particles. Most of this book is about eukaryotic cells, discoveries about how they work, and technologies we can use to manipulate them to do our bidding.

What about molecules? Molecules are atoms bonded together into stable configurations. They are all around us in the biological and non-biological worlds. Teflon and nylon are man-made molecules. *DNA* and *proteins* are examples of nature's molecules, but they too can be made by humans. DNA and proteins are at the core of life itself. DNA is the cell's hereditary material, and it carries the information needed for the cell to make proteins. Proteins are directly or indirectly responsible for virtually all of the parts of a cell and their various functions. We explore these two types of molecules further later on in this chapter.

Chemists and biologists diagram molecules to show how the atoms are bonded to each other (fig. 1.2). For example, a water molecule contains one oxygen atom (O) and two hydrogen (H) atoms, which is why we designate water as H_2O. Similarly, the blood sugar, glucose, contains six carbon (C) atoms, twelve H atoms, six O atoms and is designated $C_6H_{12}O_6$. DNA and proteins are very large molecules, containing thousands of atoms.

To sum up, atoms comprise molecules, molecules form cells, ordered arrays of cells form *tissues* and *organs,* and organ systems cooperate to sustain organisms. The most common atoms in biological molecules are C, O, and H. Other important but less common atoms in biological molecules are phosphorus (P), nitrogen (N), sulfur (S), and iron (Fe). Many trace elements are also important to cells. Some of these are in the ingredients list on your bottle of multivitamins.

What Do Cell Biologists Do?

Cell biologists are "Renaissance" scientists. They draw upon methods and findings from a variety of sources and integrate these into the relatively young discipline called cell biology. Information from genetics (study of heredity), cytology (study of cell structure), biochemistry (chemistry of life), physics (study of energy and matter), and molecular biology (study of DNA function) contributes to the cell biologist's understanding of how a cell works.

Sometimes cell biologists focus on the very small, like the shape or function of specific molecules inside a cell, and other times they take a panoramic view of cells and examine how the shapes, spatial arrangements, and communication between millions and billions of cells produce functioning tissues and organs. In the end, a cell biologist's goal is to explore the relationship between the structure and function of cells and their parts.

Cell biology emerged as a discipline during the early 1960s due to two new technologies: biological *transmission electron microscopy (TEM)* and *high-speed centrifugation*. TEM (fig. 1.3, bottom) allowed the structures of very small components of cells to be seen either for the first time or in much greater detail than previously using conventional light microscopy (fig. 1.3, top). When high school biology students use light microscopes to examine drops of pond scum for microscopic life, they see objects 500 times smaller than those visible to the unaided human eye.[2] TEM can visualize objects 250 times smaller than those seen with the best light microscope; that is,

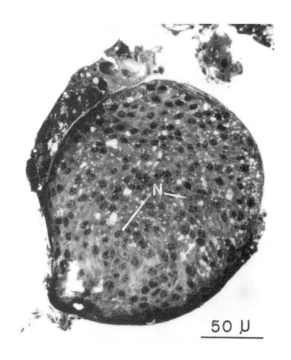

50 μ

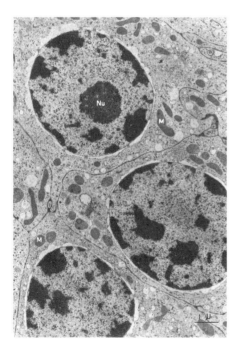

Fig. 1.3. Cell structure seen by light microscopy (LM, *top*) and transmission electron microscopy (TEM, *bottom*). Photographs are of a gland in crickets that produces a hormone controlling insect growth and development. The dozens of small roundish gray objects (N) with dark speckles in the top photo are cell nuclei. Each cell has a nucleus that contains all of the cricket's genetic information. Portions of three cells and their nuclei are shown in the bottom photo. A specialized part of the nucleus, the nucleolus (Nu) is visible in one cell and numerous mitochondria (M) appear in the cell's cytoplasm surrounding each nucleus. TEM reveals more subcellular detail than LM, but LM can show how cells are organized into tissues and organs.

TEM can show us subcellular structures whose sizes correspond roughly to the width of a DNA molecule.[3]

High-speed centrifugation of cell homogenates (mixtures of the contents of purposefully ruptured cells suspended in solution) allowed biologists to isolate purified, subcellular components called *organelles* (little organs). Soon biochemists began learning about the biochemical functions performed by each type of organelle. Discoveries made possible by TEM and high-speed centrifugation, laboratory methods still widely used, gave birth to a new discipline focused on the relationship between subcellular structure and biochemical function, cell biology.

Cell Structure and Function

Before the rise of molecular biology in the 1970s, biologists used structural criteria to classify cells as either prokaryotic or eukaryotic. Eukaryotic cells possess a true, *membrane*-bound *nucleus* (*eu* = true; *karyon* = nucleus) easily visible by light microscopy as a roughly spherical object comprising about one-third of the cell's volume. The nucleus contains DNA and an array of proteins that facilitate DNA's function as the hereditary material. The non-nuclear region of the cell is called the *cytoplasm*. Prokaryotic cells lack a true nucleus (*pro* = before; *karyon* = nucleus), so the genetic material resides in the cytoplasm along with other cellular constituents. The word *prokaryotic*, "before a nucleus," describes cells that appeared on the Earth before eukaryotic cells.

The nucleus is the most prominent structure inside eukaryotic cells. It is surrounded by two membranes that compartmentalize DNA from other cellular components. The term *membrane* refers to a thin, double-layered film of fatty molecules called *lipids,* so the double membrane surrounding the nucleus consists of two such lipid films.[4] Membrane-bound compartments in the cell separate different sets of molecules and biochemical reactions from each other. Numerous *nonmembrane-bound organelles* also populate the cytoplasm of eukaryotic cells. *Prokaryotes* and *eukaryotes* have a few nonmembrane-bound organelles in common, but most organelles are strictly eukaryotic (fig. 1.4).

How important is normal organelle function for human health? The answer comes from a brief look at what membrane-bound organelles like the *nucleus, mitochondrion,* and *lysosome* and nonmembrane-bound organelles like *microtubules, microfilaments,* and *ribosomes* (fig. 1.4) do for the eukaryotic cell and what happens when they fail.

The mitochondrion is the "powerhouse" of the cell because it is where energy-rich food molecules are burned (oxidized). Energy released by oxidizing food is used to produce *ATP,* the common energy currency for the cell. Just as the euro is the common monetary currency for European Alliance countries, ATP is the one energy-rich molecule recognized and used by all parts of the cell to maintain life's activities. Without ATP, cells die. ATP fuels cell movement, reproduction, sensory perception, and even our thoughts. Like the nucleus, mitochondria have two membranes surrounding them. Inside the mitochondrion's inner chamber and within the inner membrane itself, energy-rich molecules oxidize to form ATP (fig. 1.5). A typical cell contains several hundred mitochondria, and a human egg cell contains thousands.

Mitochondria are distinctive in two other ways: they possess their own DNA, and they are inherited from mothers. The maternal inheritance of mitochondria reflects their abundance in egg cells and absence in the portion of sperm cells that enters the egg at fertilization. By analyzing genes in mitochondrial DNA, biologists gain information about the maternal ancestry of organisms. Examining human mitochondrial DNA from indigenous populations worldwide led to the conclusion that all humans descended from one woman or a small group of women who lived in Africa about 250,000 years ago.

Several pathologies and diseases are associated with abnormal mitochondria. Ischemia from strokes and myocardial infarctions causes a rapid loss of mitochondrial function and cell death in the O_2-depleted tissues. Mitochondria in liver cells of alcoholics fuse with each other to produce dysfunctional megamitochondria. Conditions associated with mitochondrial DNA mutations or damage include blindness, deafness, seizures, infertility, muscle tremors characteristic of advanced Parkinson's disease, and premature aging.[5]

Prokaryotic Cell

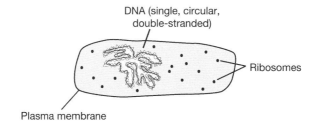

DNA (single, circular, double-stranded)

Ribosomes

Plasma membrane

Eukaryotic Cell

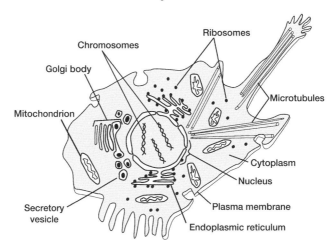

Chromosomes

Golgi body

Mitochondrion

Ribosomes

Microtubules

Cytoplasm

Nucleus

Plasma membrane

Secretory vesicle

Endoplasmic reticulum

Fig. 1.4. Prokaryotic (*top*) and eukaryotic (*bottom*) cell structure. Prokaryotic cells have no nucleus or other membrane-bound organelles, and their single chromosome consists of a circular, double-stranded DNA molecule. Eukaryotic cells have many types of membrane-bound organelles, including the endoplasmic reticulum, and various types of vesicles and the double-membrane nucleus and mitochondria. Their multiple chromosomes exist as linear molecules of double-stranded DNA. Microtubules are components of the eukaryotic cytoskeleton. Ribosomes synthesize proteins and are present in both prokaryotic and eukaryotic cells.

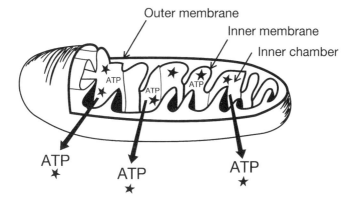

Fig. 1.5. Mitochondrion structure and function. Mitochondria are double-membrane organelles. They oxidize food molecules in their inner chamber and use the released energy to produce ATP molecules that are transported out of the mitochondrion for use by the cell. Not shown is the bacterial-like, mitochondrial chromosome, a circular, double-stranded DNA molecule within the inner chamber. Mitochondrial chromosomes are similar in structure to the prokaryotic cell chromosome depicted in figure 1.4.

Lysosomes are small membrane-bound vesicles that serve as the cell's digestive system. Inside lysosomes are powerful *enzymes* that break down large, ingested food molecules into smaller molecules that are eventually oxidized inside mitochondria or used to construct other components of the cell.[6] Lysosomes also digest worn-out organelles, recycling the breakdown products for new building projects inside the cell. Sometimes lysosomes go berserk, digesting the entire cell from the inside out. This happens in the lungs of persons afflicted with asbestosis, a disease caused by asbestos fibers entering lung cells. Inherited diseases called lysosomal storage disorders result from genetic defects in lysosomes, rendering them unable to digest certain materials. The results are an abnormal accumulation of undigested material inside the cell and devastating diseases, including Tay-Sachs disease, that cause mental retardation and joint and skeletal deformities. Fortunately, all known lysosomal storage disorders can now be diagnosed prenatally by amniocentesis, so their occurrence is declining.

Microtubules are components of the cell's *cytoskeleton,* a complex array of tubules and filaments that serve as the "bones" and "muscles" of cells. They provide support and also direct cell movement, including the movement of objects inside cells. Microtubules are thin, hollow tubes that form a cytoplasmic scaffolding that maintains cell shape and orchestrates movements like the beating of sperm tails and the separation of duplicated chromosomes before cell division. In the lungs, thousands of microtubule-containing, hair-like structures called *cilia* cover the surface of cells lining air sacs and passageways. Cilia beat constantly to provide a cleansing flow of mucus up and out of the lungs. Components of cigarette smoke poison ciliary microtubules; the cilia stop beating; nasty particles and other inhaled impurities build up in the lungs causing emphysema and cancer. Mutations in the *genes* for certain microtubule-associated proteins impair microtubule action inside sperm tails and lung cells causing infertility and emphysema.

Ribosomes are tiny protein-producing, nonmembrane-bound organelles present by the thousands in the cytoplasm of both prokaryotic and eukaryotic cells. Their job is to use genetic information in DNA to produce the thousands of different kinds of proteins required to maintain life. No cell can survive without functional ribosomes. Some antibiotics kill bacteria by specifically poisoning their ribosomes while leaving unharmed the eukaryotic ribosomes in the infected person's cells.

Next is the nucleus. Inside the nucleus are long molecules of double-stranded DNA that, along with several kinds of proteins, comprise the cell's chromosomes. The genetic information needed to make an entire individual, be it a frog, a sheep, or a human being, is contained in the DNA of the chromosomes inside the nuclei of common body cells, such as from skin or intestine. Normally, only a portion of that information is activated to direct the operations inside a particular type of cell. But, as we will see in chapter 6, on cloning, there are ways to reawaken the entirety of information stored in the DNA of a single body cell and allow it to produce a fully formed individual.

In prokaryotic cells, genetic information is carried by a single, circular, double-stranded DNA molecule (see fig. 1.4). No nucleus compartmentalizes prokaryotic DNA from the rest of the cytoplasm as in eukaryotic cells. Since bacteria are single-celled organisms, a division of labor between various cell types does not exist as it does in multicelled organisms.

Finally, how do cells sense conditions in their environment and communicate with one other? They do so by recognizing and responding appropriately to molecules in the environment and on the surfaces of other cells. Embedded in the external membrane of cells are a multitude of *receptor proteins* (receptors). The three-dimensional shapes of receptors and of signal molecules (*ligands*) in the environment and on the surface of other cells allow receptors and their ligands to fit snugly together, like the pieces of a puzzle. When receptors and ligands come together, a cascade of biochemical events appropriate for the situation ensues inside the cell. *Nerve impulses* traveling between cells (see chapter 9), responses to hormones like insulin and adrenalin, and the body's attempt to reject organ transplants are all examples of outcomes of receptor-ligand interactions.

By now you may feel overwhelmed by what goes on inside the microscopic bits of life called cells. You may even think that scientists already know nearly all there is to know about cells. First, be consoled that you don't need to know much about cells to understand the impact that biotechnologies will have on humanity, as described after this chapter. Secondly, scientists, too, are awestruck at what happens inside cells and at what they have yet to learn. The November 25, 2011, issue of *Science* magazine devotes a special section titled "Mysteries of the Cell" to articles exploring many current unknowns about the cell. Even partially solving any one of these, such as how a cell measures its size or positions certain molecules inside it, may result in future Nobel Prizes and untold numbers of PhD dissertations.

The Central Dogma of Biology

We turn now to molecular biology, a discipline devoted mainly to understanding how cells store and use genetic information. Consider DNA, the famous carrier of genetic information, and proteins, the building blocks and workhorses of cells. A concept called the central dogma of biology helps us explore the relationship between DNA, proteins, and life. The central dogma of biology is a statement about the flow of genetic information inside cells. This is an unusual use of the word *dogma* since science and dogmatic thinking are incompatible.

In 1957, at a meeting of the Society for Experimental Biology held at University College in London, Francis Crick proposed a relationship between DNA and proteins that he referred to as "the doctrine of the triad" and "the central dogma."[7] His proposal later became known as the central dogma of biology and is diagramed like this: DNA → RNA → protein.

What this means is that genetic information in DNA flows to an intermediary RNA molecule (*messenger RNA, mRNA*) and that the information in mRNA is then used to make protein. The reason this scheme is so "central" to all of biology is because DNA is the genetic material for all cells on Earth and because proteins are directly or indirectly responsible for all parts of the cells, tissues, and organs in every living thing.

The first half of the central dogma (DNA → mRNA) is called *transcription,* and the second half (mRNA → protein) is called *translation*. Important to note is that DNA does not become mRNA, and mRNA does not become protein. Rather, the arrows signify a transfer of information so that the information present in DNA (as a gene) is used to make a new molecule, mRNA, which acquires that information. The genetic information in mRNA is then used to construct a protein. At the end of the process, DNA still exists as it did at the outset, unchanged by the process of transcription, and mRNA is unchanged by the process of translation.

An Analogy

How information in DNA is used to make a human being or any other organism can be likened to how an architect's blueprints are used to construct a cathedral. To build a cathedral, the architect draws blueprints for it. Then masons, plumbers, carpenters, and electricians get copies of the blueprints to follow in the construction process. At the end of the project, the original blueprints remain safely stored in a safe box. Most of the copies end up jammed into glove boxes of workers' cars and trucks and are eventually discarded.

In this analogy, blueprints are DNA molecules, copies of the blueprints are mRNA molecules, and the cathedral is a cell or an organism built from an assemblage of proteins. The workers are ribosomes, machines that produce proteins using the instructions in mRNA. The architect is *natural selection,* the force driving most of evolution. Not included in the analogy is how mutation and the recombining of genetic material during sexual reproduction produce variations in a population of blueprints upon which natural selection acts during evolution.

To appreciate the power of some of the biotechnologies we'll be examining, such as embryo selection (chapter 3), genomics (chapters 4 and 5), genetic enhancement (chapter 6), and synthetic biology (chapter 10), we need some basic information about the three components of the central dogma: DNA, RNA, and protein. So, we look at the structure of these three types of molecules, beginning with proteins. Then we examine how genetic information flows inside a living cell, from DNA through RNA to proteins. Finally, we consider how genetic information is passed on between successive cell generations and between successive generations of individuals.

Proteins

Proteins (strings of attached *amino acids*) are among the largest molecules in the cell. They are long, linear molecules that fold and bend to acquire complicated three-dimensional shapes (fig. 1.6). One way to classify a protein is on the basis of its function: structural, catalytic, regulatory, or nutritional. Structural proteins produce structures inside the cell like microtubules or substances secreted by cells such as hair or mucus. Catalytic proteins are also called enzymes. They make it possible for the thousands of biochemical reactions inside a cell to occur at the rapid rates required to maintain life. Regulatory proteins help an organism to adjust the reactions occurring inside its cells in ways appropriate for the various internal and external conditions confronting the organism. For example, hor-

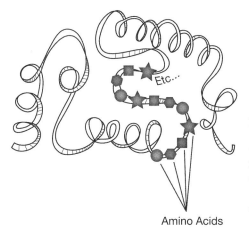

Amino Acids

Fig. 1.6. Three-dimensional protein structure. A protein's specific sequence of amino acids determines its 3-D shape, which in turn determines its function(s) in the cell. Right-handed coils, sheet-like structures, and random-appearing coils are all elements of 3-D protein structure.

mones like insulin and prolactin are regulatory proteins responding to elevated blood glucose levels and a suckling infant, respectively.[8] Other proteins called transcription factors regulate which genes are transcribed. Nutritional proteins include proteins in mother's milk and egg yolk proteins that supply nutrients to infants and to developing embryos. The different functions performed by different proteins reflect differences in their three-dimensional (3-D) shapes.

Single proteins are made of smaller subunits called amino acids, joined together in sequences specific for each protein. An average-sized protein contains about three hundred amino acids. The cell uses twenty different amino acids to build its protein chains, and it is the sequence of amino acids in a protein that determines its 3-D structure.[9] A genetic code in mRNAs specifies the sequences of amino acids in the cell's proteins. Let's see how that works.

Messenger RNA and the Genetic Code

For each protein, there is a corresponding mRNA encoding the amino acid sequence for the protein. Like proteins, mRNAs are long molecules comprised of small subunits linked together. Unlike proteins, which have twenty different kinds of subunits (amino acids), mRNAs have only four

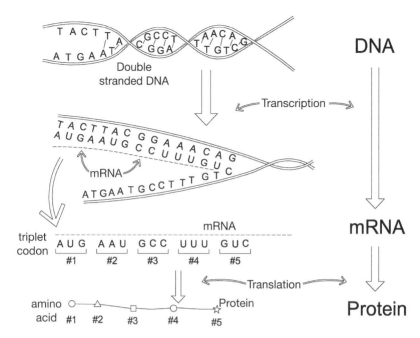

Fig. 1.7. The central dogma of biology. Double-stranded DNA opens in the region to be transcribed to produce a messenger RNA (mRNA) molecule complementary to one of the DNA strands. The mRNA moves from the nucleus into the cytoplasm, where it is translated to produce protein. Each set of three bases in the mRNA (#1–5 in the diagram) is a codon specifying a particular amino acid (#1–5) in the protein.

kinds of subunits. These are called *bases* and are designated by the letters A, G, U, and C. Bases are attached at regular intervals to a backbone of alternating sugar (ribose) and phosphate groups, depicted simply as a dotted line in figure 1.7. The sequence of bases in a mRNA determines the order of amino acids in the protein encoded by the mRNA.

How can four bases specify the order in which twenty different amino acids are joined together? Each set of three adjacent bases in a mRNA specifies one and only one of the twenty amino acids. These sets of triplet bases are called *codons,* and they constitute the genetic code. Since four different letters (A, G, U, and C) can be arranged into sixty-four (4^3) different triplets (AUG, AAC, CCG, UAC, etc.), there are more than enough triplets to code for each of the twenty amino acids. In fact, some of the "extra" codons give redundancy to the genetic code; that is, a given amino acid can be specified by more than one triplet codon. During protein formation, ribosomes move along the mRNA, translating codons as they go, and join together the specified amino acids into a chain that becomes the protein. The ribosome builds the protein one amino acid at a time in a sequence corresponding to the sequence of codons in the mRNA (fig. 1.7). So how does DNA specify mRNA molecules?

DNA and Its Transcription to Produce mRNA

DNA is a huge molecule comprised of only five kinds of atoms: carbon, oxygen, hydrogen, nitrogen, and phosphorus; but, oh, what a molecule it is! It gives continuity to all past, present, and future generations of cells and is the raw material upon which biological evolution acts. In case you were wondering, the DNA acronym stands for deoxyribonucleic acid, but everybody just says, "DNA."

Messenger RNA molecules are transcribed from segments of DNA called *genes*. DNA and RNA structures are very similar. Like RNA, DNA is a long chain of bases (A, G, T, and C) attached to a sugar-phosphate backbone. Notice that three of the four bases in DNA are also in RNA (A, G, and C) and that the U in RNA replaces the T in DNA. DNA usually occurs in a double-stranded form consisting of two complementary DNA molecules.

Rules of base pairing determine the complementarity of the two strands of a double-stranded DNA molecule. The C in one strand pairs with the G in the complementary strand, and A and T pair with each other (fig. 1.7). Each pair of complementary bases in a double-stranded molecule is called a base pair, and the length of a DNA molecule is measured in numbers of base pairs. A human egg or sperm cell contains twenty-three different DNA molecules, corresponding to its twenty-three chromosomes; together, the twenty-three chromosomes contain more than three billion base pairs of DNA. A typical gene consists of only a few thousand base pairs.

When a gene is transcribed, the complementary strands of DNA in the region of the gene's start point separate to expose the bases. Just one strand serves as a template for transcription to produce an mRNA molecule. Transcription follows the base pairing rules so that wherever C occurs in the template DNA, G is placed in the corresponding mRNA (and vice versa) and similarly for T and A, except that in RNA, the U base is used instead of T.

Applying the base pairing rules, a strand of template DNA reading TACTTACGGAAACAG will be transcribed to produce an mRNA molecule reading AUGAAUGCCUUUGUC. From left to right, this small mRNA contains five triplet codons (AUG/AAU/GCC/UUU/GUC) that specify five adjacent amino acids in a protein (fig. 1.7). In a functional RNA, the last codon is a stop codon that does not specify an amino acid but simply tells the ribosome to stop adding amino acids to the protein.[10] A protein three hundred amino acids long requires three hundred triplet codons (nine hundred bases) in its mRNA.[11]

Now we turn to chromosomes to see what happens to them when cells reproduce. We are only concerned with eukaryotic chromosomes here because the application of biotechnology to ourselves and other eukaryotes is the main concern of this book.

Chromosomes

Near the end of the nineteenth century, decades before the 1953 discovery of DNA's structure, cytologists used light microscopes to identify eukaryotic chromosomes as dark-staining bodies near the center of dividing cells. Now, from the work of individual researchers and from international endeavors such as the Human Genome Project (chapter 4) completed in the early twenty-first century, we know chromosomes much more intimately.

To appreciate how detailed knowledge about chromosomes contributes to modern biotechnologies, we must briefly consider the relationship between four genetic terms: *genome, chromosome, gene,* and *DNA.* This series represents a hierarchy of structural organization for an organism's genetic material (fig. 1.8). The genome of a plant or animal consists of the DNA in all of the chromosomes within a single cell of an organism. This DNA occurs as long, linear (as opposed to circular), double-stranded molecules

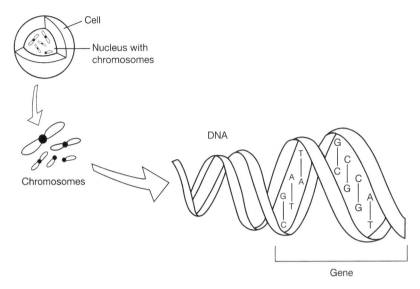

Fig. 1.8. Relationships between DNA, genes, and chromosomes. Each chromosome in the cell nucleus is a single, long, double-stranded DNA molecule containing the sequences of bases (A, G, C, T) for thousands of genes. The DNA comprising all types of chromosomes inside the nucleus of a single cell is the organism's nuclear genome.

associated with many kinds of chromosomal proteins that package DNA and regulate gene activity. Genes themselves are discrete segments of the long, chromosomal DNA molecules, and many of them carry the information needed to make proteins.

Eukaryotic chromosomes come in pairs, with one member of each pair originating from each parent. A typical human body cell has forty-six chromosomes, twenty-three from the individual's mother and twenty-three from the father.[12] A few thousand genes lie along the length of most eukaryotic chromosomes. Remarkably, more than 90 percent of the chromosomal DNA of humans and many other animals contains no protein-coding genes. This "extra" DNA is the object of intense research, and recent data indicate that much of the non-gene DNA in humans functions to regulate gene activity.[13]

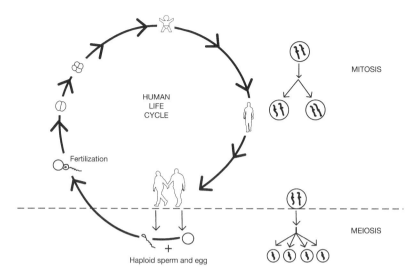

Fig. 1.9. Life cycle of humans and other sexually reproducing organisms. The fertilized egg and the cells arising from it by mitosis are diploid, having two copies of each type of chromosome. Meiosis produces haploid eggs and sperm with just one copy of each type of chromosome. Union of egg and sperm regenerates the diploid state, allowing the next generation of individuals to grow via mitosis.

Cell Division and the Chain of Life

There is an unbroken chain of life, linked by DNA replication and cell division, from the first cells some 3.8 billion years ago to us. All life forms come from ancestral life forms, and all cells come from pre-existing cells. These foundational principles of biology depend upon the replication of genetic material and cell division. Here we examine two distinct types of cell division in eukaryotic cells: *mitosis* and *meiosis*. Appreciating the differences between mitosis and meiosis is important for understanding certain biotechnologies, particularly stem cell technologies (chapter 2) and cloning (chapter 7).

Mitosis produces two new cells genetically identical to the parent cell, whereas meiosis produces cells with half the genetic content of the parent cell (fig. 1.9). Through mitosis, fertilized single eggs give rise to the trillions of cells in multicelled organisms like human beings and giant redwoods. Mitosis also regenerates tissues during our lifetimes. Blood cells, hair follicle cells, intestinal lining cells, skin cells, and many other cell types continually die, only to be renewed by the mitotic divisions of cells still living. Meiosis, by contrast, produces eggs and sperm, the basis for new generations of sexually reproducing organisms. Cells produced by mitosis have two copies of each type of chromosome and are therefore called *diploid,* whereas cells produced by meiosis (eggs and sperm) have just one copy of each chromosome and are called *haploid.*

Normal human cells have twenty-two types of chromosomes plus X and/or Y sex chromosomes. So, a diploid human cell has a total of forty-six chromosomes, and a haploid human cell has twenty-three. For the interested reader, figure 1.10 details how a diploid cell gives rise to two new diploid cells via mitosis or to four haploid cells via meiosis. Cells in figure 1.10 contain just two types of chromosomes. Because eukaryotic cells choreograph the accurate replication and correct segregation of multiple sets of chromosomes into new cells during mitosis and meiosis, we and our many cousins on the tree of life are here today.

Evolution invented sex about one billion years ago as a way to increase the genetic variation within populations, thereby making them more adaptable to changing living conditions. Meiosis occurs in the ovaries and testes and creates genetic variation for the next generation in two ways. The first is a process called *genetic recombination* or crossing over, where like chromosomes originating from the male and female parents exchange corresponding pieces of themselves so that mixtures of paternal and maternal DNA amalgamate into new, recombinant chromosomes (fig. 1.10, right). Secondly, independent and random segregation of the paternal and maternal members of each pair of chromosomes into the eggs and sperm creates a huge number of possible combinations of maternal and paternal genetic material. For example, the twenty-three pairs of chromosomes in one human cell can produce about 8 million (exactly 2^{23}) possible combinations of maternal and paternal chromosomes in eggs and sperm.[14] Since the meiotic cell in figure 1.10 (right) has just two pairs of chromosomes, the possible combinations of paternal and maternal chromosomes in eggs and sperm are

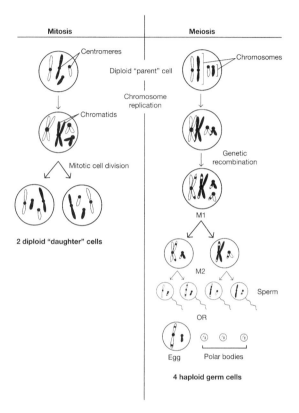

Fig. 1.10. Comparison of mitosis and meiosis. Both mitosis and meiosis begin with a diploid cell undergoing chromosome (DNA) replication. In mitosis (*left*), the two chromatids of each replicated chromosome separate during cell division so that each daughter cell has a diploid complement of chromosomes like the original parent cell. In meiosis (*right*), replicated chromosomes exchange genetic material during genetic recombination. During the first meiotic division (M1), chromatids remain unseparated and one version of each replicated chromosome goes to each daughter cell. During the second meiotic division (M2), chromatids separate and cell divisions produce four haploid germ cells, either four small, functional sperm cells or one large, functional egg and three polar bodies that eventually degenerate.

only four (2^2). In males, all four cells produced by the two meiotic cell divisions go on to form sperm. In females, only one of the four cells develops into an egg. The other three cells are called polar bodies and ultimately degenerate and disappear (fig. 1.10, right).[15]

In summary, sexual reproduction via the union of egg and sperm functions to provide the next generation with a diploid set of chromosomes in a cell that goes on to generate a new individual by mitosis. That cell is the fertilized egg. Sexual reproduction also supplies new generations of organisms with the genetic variation needed for evolution via natural selection.

The First Cells

It is one thing to imagine the trillions of cells in one's body all originating with a single, fertilized egg cell, but it is quite another to ask where the first cells on Earth came from. Let's take a brief look at what science has to say about the origin of life on Earth.

An awe-inspiring exercise is to imagine the vastness of the time that life has adorned Earth. Consider a pile of the 2,102-page *Unabridged Random House Dictionary of the English Language.* The pile is 238 feet high (without covers), as tall as a twenty-story building. Let that pile of pages represent life's 3.8 billion-year sojourn on Earth. What portion of this enormous pile of pages represents the time elapsed since early modern humans painted Paleolithic images on the walls of Lascaux cave in southern France? The answer: one two-sided page! On that single page is written humankind's invention of agriculture; the origin of the world's great religions; the rise and fall of great civilizations in Egypt, China, India, Europe, and the New World; the rise and fall of Islamic science; the Western world's Middle Ages, Renaissance, Scientific Revolution, Enlightenment, and Industrial Revolution; two world wars; and revolutions in physics and biology. Unwritten in the fossil record, someplace before page 1 of that twenty-story-high pile of pages, lies the dawn of life.

Detailed discussion of science's attempts to understand how the first cells emerged from nonliving matter on Earth is beyond the objective of this book, but several excellent and accessible works exist on the subject.[16]

Here we briefly consider some aspects of life's antiquity and origins upon which virtually all biologists agree. Information about natural selection's creativity helps to give scientific and ethical perspective to humankind's increasing power over its own future evolution via embryo selection (chapter 3) and genetic enhancements (chapter 6) and to the human design and creation life (chapter 10).

First, consider the origin of the elemental building blocks for life: carbon (C), oxygen (O), nitrogen (N), hydrogen (H), sulfur (S), phosphorus (P), iron (Fe), and several trace elements. These and all other elements found on Earth were forged in the core of a massive, collapsing star in our galactic neighborhood during the final stages of its life. Spewed out into space when the dying star exploded as a super nova, this elemental stardust is the substance that formed the planets and other bodies that coalesced to create our solar system about five billion years ago. Since all of the elements in living cells come from the planet Earth, we and all other living things in the biosphere are stardust, literally. What a wonderful thing for us all to have in common.

It took about 1 billion years for prokaryotic (bacterial) life to emerge on the surface of the new Earth. Thinly sliced rocks dated 3.5 to 3.8 billion years old and examined microscopically contain fossilized remains of prokaryotic cells. These cells were non-photosynthetic. *Photosynthesis* releases oxygen (O_2) into the atmosphere, and geological evidence shows that oxygen did not start accumulating in the atmosphere until about 2.4 billion years ago.

Exactly when eukaryotic cells first appeared on Earth is controversial. Fossil evidence for cells containing a membrane-bound nucleus, the hallmark of eukaryotic cells, is present in rocks about 2 billion years old. But from other evidence, most biologists believe that the eukaryotic lineage is older than that, perhaps as ancient as 3 billion years.[17] Although prokaryotic cells sometimes grow in colonies, true multicellular organisms are all eukaryotic. Fossil evidence for the oldest known multicellular organisms appears in rocks 2.1 billion years old (Donoghue and Antcliffe 2010; El Alani et al. 2010).

Endosymbiotic Model for the Orgin of Mitochondria

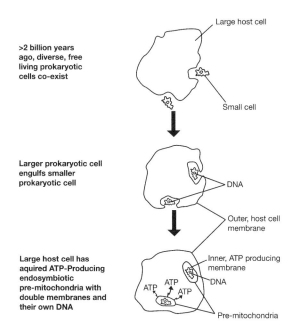

Fig. 1.11. The endosymbiotic origin of mitochondria. A small prokaryotic cell with
a circular chromosomal DNA and ATP-producing enzymes in its outer membrane
established an obligatory endosymbiotic relationship with a larger prokaryotic cell that
lacked an ATP-producing membrane. Engulfment of the smaller cell gave the larger cell
an additional, host-derived membrane that became the outer membrane of modern-day
mitochondria.

How eukaryotic cells originated is better understood than exactly when
they first appeared. There is widespread agreement among molecular biologists, microbiologists, and cell biologists that eukaryotic cells arose when
some small prokaryotic cells took up permanent residence inside a larger
pre-eukaryotic cell. When one cell lives inside another cell, the relationship is called endosymbiosis, *symbiosis* being the living together of two different organisms in a mutually beneficial association. An endosymbiotic
model for the origin of eukaryotic cells was proposed in 1970 by Lynn
Margulis. Today Margulis's model enjoys overwhelming supportive evi-

dence and is widely accepted as the best explanation for the origin of eukaryotic cells.

According to the endosymbiotic model, the mitochondria inside today's eukaryotic cells arose when prokaryotic cells efficient at making ATP (aerobic cells) became engulfed by the *plasma membrane* of larger host cells (anaerobic cells) that lacked an efficient way to make ATP (fig. 1.11). When the *symbionts* began living inside the larger host cells, both types of cells gained something from the new association. Host cells gained ATP from the symbionts, and symbionts gained protection and the stability of a controlled environment in which to live from the host cytoplasm. This scenario explains the structure and function of today's mitochondrion—its double membrane, circular DNA molecule, and inner membrane that produces ATP. The structure and function of chloroplasts—double-membraned, DNA-containing organelles responsible for photosynthesis inside plant cells—are similarly explained. For me, it is profound to realize that every cell in my body (and in the body of every other eukaryotic organism) is actually a community of organisms working together to maintain the life of a vast assemblage of communities that I call myself.

The Unity of Life

Just as interesting as the origin of cells is what the biology of cells reveals about the unity of life. By "unity," I mean shared likenesses. The most significant of life's shared likenesses are the common ancestry and the interdependent ecological relationships among all living things. Formed over eons of biological evolution, common ancestry and ecological interdependencies connect all members of the living community to each other. A common ancestry for all living things is reflected in the structures of genes and proteins and their genetic and metabolic functions (Steele and Penny 2010; Theobald 2010). Ecological interdependencies are all around us—from animals inhaling life-sustaining oxygen produced by green plants to flowering plants' reliance on insects and bats for their pollination. This tightly woven web of inherited and ecological connections is the fundamental fabric of nature.

Cells and Human Values

What does modern cell biology have to do with human values? As we move through the topics in this book, we will see that modern biotechnologies raise questions and issues about life that were less urgent or nonexistent before the Age of Biotechnology. Some of these include defining personhood and the moral status of human embryos, describing human nature and choosing whether or how to change it via biotechnology, defining life, learning how to behave as creators of new life forms, and deciding what elements of life may be sold or patented. Rebecca Skloot (2010) treats the latter issue especially poignantly in her 2010 book about Henrietta Lacks, a woman who, without giving her consent, supplied cells used for human cell research and for profit worldwide for the past six decades. Knowing about cells, genes, and embryos is important for tackling the tough ethical issues raised by modern biotechnologies.

Science cannot tell us how to proceed in these and other areas where biology, ethics, and society intersect. For decision-making guidance on the future directions for biotechnology research and use, science needs to hear from non-scientists and, especially, from the lay public.[18] In democratic societies, citizens ought to have a significant voice in how new technologies are developed and used. So we all must decide where to stand on the ethical issues raised by twenty-first-century biotechnologies and speak out with our votes, letters, and personal interactions. There is no reversing the continuing advancements in biotechnologies. And as time passes, opportunities to influence how biotechnologies are woven into the fabric of human society may dwindle. The stakes are high. Technological and intellectual revolutions in the past have wrought permanent changes in how we live, behave, and think, and the biotechnology revolution will do the same.

For the lay public to provide wise guidance to scientists and legislators, it must know something about the science upon which society-altering biotechnologies are based. This is why knowing about cells matters. This book aims to encourage informed decision-making that is respectful of and empathetic with others' views. Although increased scientific literacy may encourage convergence of opinion on certain ethical issues, consensus on any one issue is unlikely in pluralistic societies. Nevertheless, knowing that persons with views different from one's own consult accurate scientific information to form those views should make it easier to respect and empathize with them.

Conclusion

Cells are the basic units and DNA is the genetic material for all living things on Earth. All cells use the same genetic code to specify the structure of proteins, and this reflects a common ancestry for all of Earth's life forms that extends back in time for nearly four billion years. Molecular biology emerged as a discipline following the 1953 discovery of the structure of DNA, and cell biology became a discipline in the 1960s due to technological breakthroughs for studying relationships between cell structure and function. These two disciplines are at the core of modern biotechnologies that challenge our notions about human nature and the morality of redesigning life, including ourselves.

Questions for Thought and Discussion

1. Recall the cathedral analogy used to describe the flow of genetic information via the central dogma. How could you extend this analogy to accommodate evolution, including mutations and natural selection? How do you interpret the architect metaphor in the analogy?
2. What "ways of knowing" are there besides science? How does science as a way of knowing differ from these others?
3. What effect do you believe the discovery of extraterrestrial life (either intelligent or non-intelligent) would have on human values? How similar chemically do you think extraterrestrial life would be to life on Earth?
4. Do you agree with the author that knowing basic information about cells matters? If not, why not? If so, why?
5. How have discoveries in cell biology affected your life?

Notes

1. A scientific theory is the closest thing to "truth" that science offers. Science is non-dogmatic and self-correcting, so it never claims to own the absolute truth. Hypotheses about the physical world are tested with experiments designed to invalidate the hypotheses if the hypotheses are wrong. After many experimental tests by many persons using many different approaches, a hypothesis may be elevated to the level of theory. Scientists feel quite confident about the ability of theories to explain reality, but they are always open to new data that require discarding a theory, revising it, or replacing it with a new theory. Albert Einstein's theory of general relativity replaced the classical Newtonian theory of physics, and Copernicus's heliocentric (sun-centered) theory replaced Aristotle's and Ptolemy's geocentric (Earth-centered) theory of the universe. The germ theory of disease posits that bacteria and viruses cause infectious diseases, and the theory of biological evolution posits that different life forms on Earth have a common ancestry.

2. Unaided human eyes see objects as small as 0.1 millimeters (0.0036 inches) in diameter.

3. TEM discerns smaller objects than light microscopy because the electron beams used in TEM have a shorter wavelength than the visible light used in light microscopes. The shorter the wavelength, the finer is the detail that can be seen.

4. Lipids comprise a broad group of naturally occurring water-insoluble molecules including fats (triglycerides), waxes, and sterols like cholesterol. Their functions in cells include energy storage and membrane formation. The major class of membrane lipids is phospholipids. One end of a phospholipid is water-insoluble and the other end is water-soluble. In an aqueous environment, phospholipids naturally form lipid bilayers, with their water-soluble ends facing outward.

5. Research with mice shows that experimentally induced mutations in mitochondrial DNA cause diverse symptoms of premature aging, including graying hair, hearing loss, and osteoporosis. But these results do not necessarily mean that damage to mitochondrial DNA causes the normal aging process. For further information on this and other pathologies associated with dysfunctional mitochondria see Karp 2010.

6. Oxidation is the chemical equivalent of slow, controlled burning to release energy that the cell can use for life processes.

7. Francis Crick's musings in preparation for his 1957 talk are preserved in notes he made to himself in October 1956. These are available to view on the National Institutes of Health Profiles in Science site: http://profiles.nlm.nih .gov/SC/B/B/F/T/_/scbbft.pdf (accessed July 23, 2009). Years later, in his 1990 book, *What Mad Pursuit: A Personal View of Scientific Discovery,* Crick explains that he came to use the word *dogma* simply to describe "a grand hypothesis that, however plausible, had little direct experimental support" (1990, 109) at the time.

8. Insulin facilitates the conversion of glucose to the energy storage molecule, glycogen, in the liver. Prolactin released from the pituitary gland in response to stimulation of the nipple by a suckling infant facilitates the production of breast milk.

9. Although the genetic code specifies twenty different amino acids, all twenty of them are not necessarily used in every protein.

10. There are three stop codons: UAG, UGA, UAA.

11. Actually, pre-mRNAs in eukaryotic cells are about five times as long as their corresponding mRNAs because many of the bases in transcribed genes do not code for amino acids. These noncoding regions of genes are called *introns,* while the coding regions are called *exons.* The transcribed introns in premRNA must be discarded and the exons joined together to produce a functional mRNA.

12. Humans have 22 types of so-called autosomal chromosomes (autosomes) and 2 types of sex chromosomes, X and Y. Most body cells therefore have 44 autosomal chromosomes and 2 sex chromosomes because the fertilized egg that gave rise to the cells had 22 autosomes and 1 sex chromosome of its own, and so did the fertilizing sperm cell. In females, both sex chromosomes in a body cell are X, whereas male body cells have X and Y chromosomes. All eggs carry a single X chromosome and 22 autosomes. On the other hand, a sperm cell may carry either one X or one Y chromosome, along with the 22 autosomes. Therefore, the sex of a child is determined by the sperm that fertilizes the egg. A Y-carrying sperm produces a boy, while an X-carrying sperm produces a girl. One exception to the complement of 46 chromosomes in body cells are some liver cells that replicate

their chromosomes once or twice without subsequently undergoing cell division. In such cases, the cell ends up with 92 or 184 chromosomes, respectively.

13. The most common way that non-gene DNA regulates the activity of genes is by serving as a binding site for proteins that turn on or turn off the activity of certain genes. Elucidating the function of non-gene, human DNA is an objective of the ENCODE Project: http://www.genome.gov/10005107 (accessed November 11, 2012).

14. To avoid confusion, I have written that humans have 23 pairs of chromosomes, 22 pairs of autosomal chromosomes and 1 pair of sex chromosomes. But since there are two types of sex chromosomes, X and Y, humankind as a species actually has 24 types of chromosomes. Male body cells have both X and Y chromosomes, so they contain 24 types of chromosomes; human female cells lack the Y chromosome and have 23 types of chromosomes.

15. The first meiotic division is called M1 and the second division is called M2. During M1, chromatids remain unseparated and one version of each replicated chromosome enters each daughter cell. During M2, chromatids separate and cell divisions produce four haploid germ cells, either four small, functional sperm cells or one large, functional egg and polar bodies that eventually degenerate. Polar bodies are so named because they usually remain associated at one "pole" of the larger cell after cell division. Not depicted in figure 1.10 is the fact that during egg formation the first polar body forms during M1 and another one forms when the future egg cell divides again during M2. Sometimes the first polar body does not divide again, in which case only two polar bodies are present at the end of M2. But if the first polar body does divide, there will be three polar bodies at the end of M2.

16. Here are three excellent books on the scientific approach to understanding how life may have originated on Earth: Frye 2000 (*The Emergence of Life on Earth: A Historical and Scientific Overview*); Hazen 2005 (*Genesis: The Scientific Quest for Life's Origins*); and Jastrow and Rampino 2008 (*Origins of Life in the Universe*).

17. "When did eukaryotic cells (cells with nuclei and other internal organelles) first evolve? What do we know about how they evolved from earlier life forms?" October 21, 1991. http://www.scientificamerican.com/article.cfm?id=when-did-eukaryotic-cells&page=2 (accessed July 15, 2010).

18. There is no better source for this notion than Thomas Jefferson, who wrote, "An enlightened citizenry is indispensible for the proper functioning of a republic." http://www.famguardian.org/Subjects/Politics/ThomasJefferson /jeff1350.htm (accessed May 30, 2012). Since biotechnology affects individuals and society, and since individuals and society are cornerstones of our republic, we all must strive to be well informed about biotechnology for the good of the republic, as well as for our own well-being.

Sources for Additional Information

Cudmore, L.L.L. 1977. *The Center of Life: A Natural History of the Cell.* New York: Quadrangle Books. An overview of how cells work, written by an English professor enthralled with cell biology.

Dawkins, R., 2009. *The Greatest Show on Earth.* New York: Simon & Schuster. A lucid, accurate, and color-illustrated book about evolution.

DeDuve, C. 1991. *Blueprint for a Cell: The Nature and Origin of Life.* Burlington, NC: Neil Patterson Publ. Beautifully written and illustrated overview of cells, genes, and the cellular basis for life.

Ford, N. M. 1988. *When Did I Begin? Conception of the Human Individual in History, Philosophy, and Science.* Cambridge, UK: Cambridge University Press. An excellent, well-illustrated overview of human embryonic development and the beginning of personhood.

Frye, I. 2000. *The Emergence of Life on Earth: A Historical and Scientific Overview.* Brunswick, NJ: Rutgers University Press. An accessible and critical overview of science's approach to the question of how life first began.

Gibson, G., and S. V. Muse. 2002. *A Primer of Genome Science.* Sunderland, MA: Sinauer Associates. An excellent source for basic information about the central dogma of biology.

Karp, G. 2010. *Cell and Molecular Biology: Concepts and Experiments.* 6th ed. Hoboken, NJ: Wiley & Sons. A first-rate, well-illustrated cell biology textbook.

Mercer, E. H. 1961. *Cells and Cell Structure.* London: Hutchinson Educational. This book inspired me to become a cell biologist when I was in eighth grade.

Moore, J. 1999. *Science as a Way of Knowing: The Foundations of Modern Biology.* Cambridge, MA: Harvard University Press. An excellent place to learn about what science is, how it is done, and what it is not.

Rensberger, B. 1996. *Life Itself: Exploring the Realm of the Living Cell.* New York: Oxford University Press. Nicely illustrated work on cell biology.

Ridley, M. 1999. *Genome: The Autobiography of a Species in 23 Chapters.* New York: HarperCollins Publ. A *New York Times* best seller on the Human Genome Project and what it accomplished.

Science News Focus. 2011. "Mysteries of the Cell." *Science* 334: 1046–51. Short essays, each by a different author, describe five of the many major problems remaining to be solved in cell biology.

Shubin, N. 2008. *Your Inner Fish: A Journey into the 3.5-Billion Year History of the Human Body.* New York: Random House. An excellent place for non-biologists to learn about evolution.

Thomas, L. 1974. T*he Lives of a Cell: Notes of a Biology Watcher.* Toronto: Bantom Books. A classic collection of essays about cells and how they work, along with some philosophical musings inspired by cells.

Wilson, Edmund B. 1925. *The Cell in Development and Heredity (Genes, Cells, and Organisms).* New York: Macmillan. Wilson (1856–1939) is credited with being America's first cell biologist.

2
Stem Cells

Embryos, Therapeutic Cloning, and Personhood

It is doubtful that natural sexual reproduction, with its risk of sexually transmitted disease, its high abnormality rate in the resulting children, and its gross inefficiency in terms of the death and destruction of embryos, would ever have been approved by regulatory bodies if it had been invented as a reproductive technology rather than simply "found" as part of our evolved biology.

—John Harris, *Enhancing Evolution:*
The Ethical Case for Making Better People

Not since Copernicus and Galileo removed Earth from the center of the universe and Charles Darwin placed the origin of humans and other animals on equal footing has a scientific discovery created more political, religious, and social controversy than the 1998 generation of human *embryonic stem cells (ESCs).*

In that year James Thomson and his coworkers at the University of Wisconsin described how they used very early human embryos to obtain cells that could revolutionize medicine. Thomson's work was honored in 1999 by the American Association for the Advancement of Science as the Breakthrough of the Year, "a rare discovery that profoundly changes the practice or interpretation of science or its implications for society" (1999, 2238). Since 1998, cell biologists working with human ESCs in laboratories worldwide have laid the groundwork for *regenerative medicine,* a new field that will someday cure now incurable diseases, repair tissue damaged by strokes, heart attacks, and accidents, and offer humans longer, more active lives. This chapter is about the science of stem cells, the new medical treatments they promise, and the ethical dilemmas they raise.

Three categories of stem cells occur naturally in humans and other mammals: ESCs, *adult stem cells,* and *embryonic germ cells.* In this chapter we examine each of these stem cell types for their biology and clinical potential and for the ethical issues their use raises. We also look at *induced pluripotent stem (iPS) cells,* cells that do not occur naturally but are products of the laboratory. These iPS and RiPS cells shook up the world of stem cell research in late 2007 and 2010, respectively, because they behave very much like ESCs and do not involve destroying human embryos. As we will see though, iPS cells carry ethical issues of their own.

The goal of this chapter is to facilitate the formation of informed opinions and decisions about stem cells and their use. To this end, we consider six questions:

1. Where do ESCs, adult stem cells, embryonic germ cells, and iPS cells come from and what do they do?
2. What are the potential health benefits from each type of stem cell?
3. What is *therapeutic cloning,* how is it related to ESC research, and why do it?
4. What ethical issues are raised by stem cell research, the clinical use of stem cells, and therapeutic cloning?
5. How do persons from different religious traditions and cultures and ethicists view the moral status of human embryos?
6. How should benefits from new biotechnologies be distributed and biotechnology research be prioritized?

Stem Cells: The Biology

Occurrence, Normal Functions, and Clinical Uses of Stem Cells

Stem cells get their name from the fact that they can generate specialized types of cells, just as the growing tip of a plant stem gives rise to root, branch, leaf, and flower cells. In fact, some stem cells can generate all 252

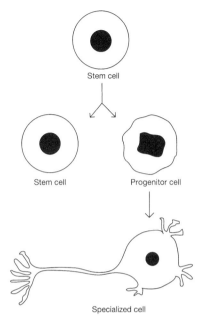

Fig. 2.1. Generalized stem cell biology. A stem cell divides to produce another stem cell like itself and a progenitor cell that goes on to develop into a specialized cell type.

types of cells in the human body. Ancestral lines for the trillions of cells in the human body originate in a small group of *pluripotent* (having a plurality of *developmental potentials*) stem cells present in very young embryos and, ultimately, from the fertilized egg cell itself.

In most cells, mitotic cell division produces two identical daughter cells. But when a stem cell divides, it produces two kinds of cells: (1) another cell like itself and (2) a *progenitor cell* committed to producing specialized cells via subsequent cell divisions (fig. 2.1). The stepwise specialization accompanying cell division that produces cells dedicated to specific tasks is called *cell differentiation*. Examples of differentiated cells include oxygen-carrying red blood cells, electrical signal-conducting nerve cells, contracting muscle cells, and insulin-producing pancreatic cells. Development of an entire plant or animal from a single fertilized egg, wound healing, and the daily replacement of worn-out blood and skin cells are examples of what stem cell division and subsequent cell differentiation can accomplish.

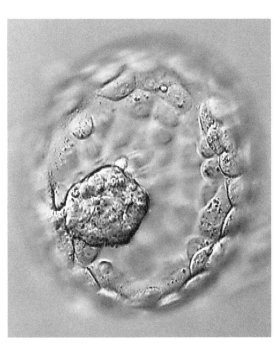

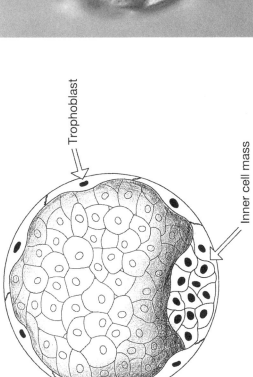

Fig. 2.2. *Left*, blastocyst-stage embryo characteristic of humans and other mammals, showing its two components, the trophectoderm, comprised of trophoblast cells, and the inner cell mass. In blastocysts implanted in the uterine wall, the trophectoderm forms the fetal portion of the placenta, and the inner cell mass eventually develops into the fetus. *Right*, four- to five-day-old human blastocyst. Photograph by Eve Feinberg, Fertility Centers of Illinois, Highland Park, IL, used with permission.

Trophoblast

Inner cell mass

To date, of the four major types of mammalian stem cells—ESCs, adult stem cells, embryonic germ cells, and iPS cells—ESCs show the greatest potential to relieve human suffering. Yet ESC research provokes heated dispute in churches, Congress, classrooms, public media, bioethics committees, political campaigns, homes, and the streets. Alas, the passion in ESC debates is matched by public misunderstandings about what ESCs are and where they come from. Therefore, let's consider ESCs first—their source, biological properties, and clinical potential.

Embryonic Stem Cells. Human ESCs originate from pluripotent cells in five- to seven-day-old embryos called *blastocysts.* At this age, the embryo is about 0.2 millimeters in diameter, less than one-fifth the size of a pinhead. The blastocyst is a hollow ball of cells whose interior contains a small group of undifferentiated cells, the *inner cell mass,* growing against the inside wall of the embryo (fig. 2.2). In a five-day-old blastocyst, the inner cell mass contains about one hundred cells. All of the cells look alike; none of them are specialized as heart, muscle, liver, brain, or other cell types. Laboratory-cultured ESCs are derived from the cells of the blastocyst's inner cell mass.

All blastocysts used for ESC research come from *surplus embryos* stored frozen at *in vitro fertilization (IVF)* clinics where couples go for assisted reproduction. Couples using IVF to have a child generally produce one to two dozen embryos. Physicians usually transfer only three or fewer embryos at a time to the woman's womb. Surplus embryos are blastocyst or earlier-stage embryos left over after a couple has completed the IVF procedure (fig. 2.3). If a couple no longer wishes to keep their embryos in frozen storage, they consent for them to be discarded, donated to infertile couples, or donated for research. At any one time over four hundred thousand surplus embryos exist in the United States alone. The majority of these are eventually discarded.

Biologists receiving surplus embryos for ESC research first thaw and reanimate them in glass dishes. When an embryo is at the blastocyst stage, researchers remove the inner cell mass and disperse its pluripotent cells into a liquid nutrient solution in a laboratory culture dish. These ESCs remain

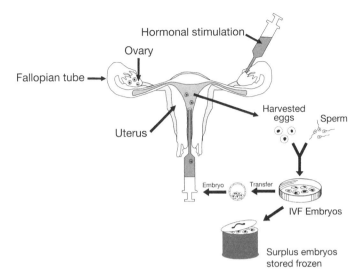

Fig. 2.3. Assisted reproduction by in vitro fertilization. Superovulation of the woman's ovaries by hormone injection causes release of up to two dozen eggs all at once. These are collected and fertilized in the laboratory to generate many blastocyst embryos. Two or three embryos are typically transferred to the uterus. The remaining embryos are stored frozen in liquid nitrogen for later attempts at implantation.

undifferentiated until stimulated with hormonal or chemical agents that induce them to produce specialized cell types (fig. 2.4). Some people believe that ESCs come from the wombs of pregnant women, umbilical cord blood, or aborted fetuses. Mistaken notions like these may contribute to some persons' vehement opposition to ESC research. In fact, surplus embryos used for the production of ESCs never would have become implanted in a woman's uterus.

The pluripotency of ESCs is what makes them so valuable for regenerative medicine. Eventual application of this research to repair diseased, injured, and worn-out tissues could improve, save, and lengthen the lives of millions of people. Diseases likely to respond to ESC therapy include insulin-dependent diabetes, Parkinson's disease, chronic heart disease, liver failure, kidney disease, multiple sclerosis, osteoarthritis, rheumatoid arthritis, some cancers, and even Alzheimer's disease. ESCs could become

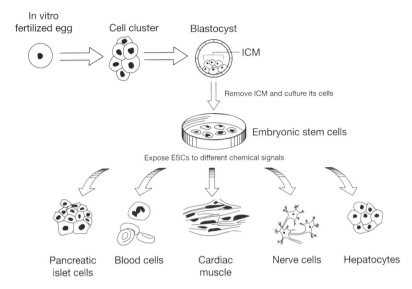

Fig. 2.4. Creation of an embryonic stem cell (ESC) line from cells of the inner cell mass (ICM) of a blastocyst-stage embryo. Pluripotent ESCs grown in the laboratory are chemically stimulated to develop into the desired, specialized cell types. Current research aims to develop ESC therapies to repair diseased and injured tissues.

the preferred treatment for heart, kidney, and liver disease, replacing organ transplantation as treatment for these conditions. Skin and other tissues destroyed by severe burns could also be repaired using ESC therapy.

Over 250,000 Americans suffer from spinal cord injuries, with about eleven thousand new injuries occurring every year. Nearly 60 percent of all victims of spinal cord injuries are less than thirty years old. Research during the past decade or so gives hope that ESC therapy may soon be able to return normal life to new victims of spinal cord injuries. In 1999, researchers at the Washington School of Medicine in St. Louis, Missouri, used mouse ESCs to restore nearly normal function to the paralyzed legs of rats whose spinal cord had been traumatically injured. In 2005, workers at the University of California, Irvine, used human ESC-derived cells to restore mobility to rats nearly paralyzed by spinal cord injuries. In January 2009, the Food and Drug Administration (FDA) approved a clinical trial of this treatment in human subjects with spinal cord injuries (Pollack 2009).[1] This phase 1 trial to test the safety of the procedure is being conducted by the California biotech company Geron on persons with injuries seven to fourteen days old. The reason for restricting the trial to very recently injured person is that rodent studies indicate that only recently injured animals respond to ESC treatment. The FDA placed a short hold on initiating the trial due to the finding that the treatment induces non-malignant cysts in some rats injected with the stem cells. This problem was solved, and the hold was lifted in July 2010.[2] An encouraging testimonial was received in April 2011. The first patient to receive the treatment, a twenty-one-year-old Alabama man partially paralyzed in a car crash, reports renewed sensation in his paralyzed legs (Brennan 2011). A stem cell cure for spinal cord injuries may still be some years away, but positive results from this trial will be a big step toward that goal.

Finding ways to regenerate healthy tissue is not the only benefit of ESC research. Drug discovery, toxicity testing, and basic research into birth defects are other active ESC research areas. New drug development and research on the harmful effects of environmental pollutants and other toxins have relied heavily on animal testing. This situation leaves specific knowledge about the harmful side effects of drugs or the dangers of other toxins unknown until people are actually exposed to the substances and can also pose animal rights problems. Animal testing does give information about the effects of chemicals on whole organisms. But performing drug and toxicity tests on specialized cells derived from human ESCs gives researchers valuable information about chemical effects on an array of human cell types under controlled conditions, information that is otherwise difficult or impossible to obtain. Finally, insight into causes and prevention of birth defects may be obtained by studying ESCs carrying the genetic defects that cause those abnormalities.

Adult stem cells. From fetal life until death, cells wear out and die. Replacing dead cells is the function of adult stem cells that populate most tissues and retain the ability to divide. The developmental potential of adult

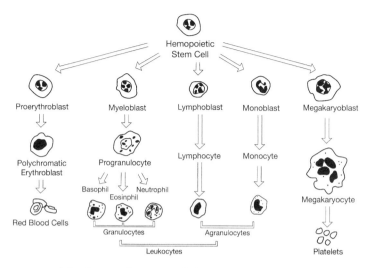

Fig. 2.5. Normal developmental potential of a hemopoietic stem cell (HSC). HSCs in the bone marrow or in umbilical cord blood are a type of adult stem cell. They normally give rise to red blood cells, all of the cellular immune system (white blood cells or leukocytes), and platelets, cell fragments involved in blood clotting.

stem cells is much more restricted than for ESCs. Normally, adult stem cells only give rise to the cell types in the tissue where they reside, so they are called *multipotent* instead of pluripotent.

The best-studied adult stem cell is the hemopoietic stem cell (HSC) that replenishes the seven types of specialized hematic cells that populate the blood and lymphatic systems (fig. 2.5). Hematic cells include (1) red blood cells that carry oxygen throughout the body; (2) five kinds of white blood cells that fight viral, bacterial, and parasitic infections; and (3) the megakaryocyte, a colossal cell that fragments into "minicells" called platelets, required for blood clotting. HSCs are concentrated in the bone marrow and in blood from the umbilical cord and placenta, but even in these tissues they account for only about one in ten thousand cells. Some scientists have tried to coax HSCs into becoming pluripotent like ESCs without much success.

In addition to their restricted developmental potential, other factors make adult stem cells less attractive than ESCs for regenerative medicine research. Adult stem cells are difficult to isolate because they are intermingled with large numbers of other cell types. Secondly, when grown outside the body, adult stem cells usually die after a few divisions and spontaneously differentiate into the specialized cell types of the tissues from which they came.

Despite the difficulties in working with adult stem cells, researchers continue efforts to improve harvesting techniques and to expand the developmental potential of adult stem cells grown in the laboratory. These labors are producing some progress for stem cells obtained from umbilical cord and menstrual blood; placental, fat, neural, lung, mammary, bone marrow, and testicular tissue; and the lining of the nasal cavity. Recently, researchers in Germany and South Korea made adult stem cells from the human nervous system pluripotent by modifying them genetically, work considered later in this chapter.

Adult stem cells have helped persons with injuries and diseases for many years. For example, when patients receive skin grafts, they rely on adult stem cells in the healthy skin tissue to help repair the injury. When cancer patients receive bone marrow transplants after chemotherapy, they depend on hemopoietic stem cells to replace blood-generating cells killed by the chemotherapeutic drugs.

Even though adult stem cells may not have the clinical value of ESCs, continued research on adult stem cells is desirable for at least three reasons. First, it could lead to regenerative treatments for specific organs. For example, adult stem cell–derived progenitor cells from bladder tissue have already been used to grow functional, new bladders; and in 2008 an international team of doctors and researchers from Spain, Italy, and England used a patient's bone marrow cells to provide her with a new windpipe. Second, adult stem cells do not carry the controversial ethical issues that ESCs do since they can be harvested from tissue biopsies or recently deceased persons. No human embryos are needed to obtain adult stem cells.

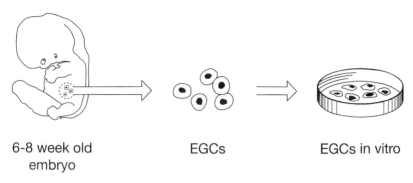

6-8 week old
embryo

EGCs

EGCs in vitro

Fig. 2.6. Derivation of pluripotent, human embryonic germ cells (EGCs) from primordial germ cells in the gonadal ridge of a six- to eight-week-old fetus obtained from an elective therapeutic abortion or a miscarriage.

Finally, there is evidence that certain cancers may arise from small numbers of adult stem cells whose mechanism for regulating cell division has gone awry, so basic research on adult stem cells could contribute to the development of new strategies for combating cancer.

Embryonic germ cells. Like ESCs, human embryonic germ cells (EGCs) are pluripotent. Unlike ESCs, EGCs are derived from five- to ten-week-old embryos/fetuses aborted spontaneously or to protect the mother's health.[3] EGCs come from a patch of tissue that contains primordial germ cells, cells that will ultimately develop into eggs or sperm (fig. 2.6). In 1998, John Gearhart and his coworkers at Johns Hopkins School of Medicine and other institutions reported growing pluripotent human EGCs in the laboratory. The cells used in the work came from embryos donated for research after therapeutic abortions and after the research protocol was approved by a clinical ethics oversight committee at Johns Hopkins University. Gearhart's results showed that the developmental potential of human EGCs equals that of human ESCs.

Despite their pluripotency, EGCs are not big players in regenerative medicine research because they must be obtained from fetal tissue, and in most countries there are stringent restrictions on using fetuses for re-

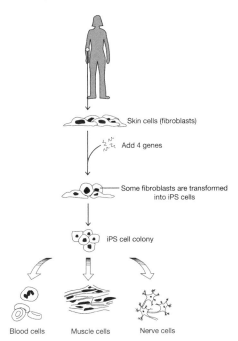

Fig. 2.7. Induced pluripotent stem (iPS) cells created by adding four genes to the genome of skin cells called fibroblasts. The added genes include transcription factors that activate numerous other genes. Although only a small fraction of treated fibroblasts develop into iPS cells, the iPS cells that are produced appear to have the developmental potential of normal embryonic stem cells.

search. Also, since EGCs are obtained from fetuses spontaneously or therapeutically aborted, the cells have little therapeutic value because they are likely to carry genetic defects that caused early termination of development in the first place. In summary, EGCs have no advantage over ESCs in regenerative medicine, so they are rarely used in current medical research.

Induced pluripotent stem (iPS) cells. As noted earlier, a remarkable breakthrough in human stem cell research occurred in 2007 when two groups of researchers independently reported that a simple genetic manipulation of human skin cells reprograms them to become pluripotent cells called iPS cells (fig. 2.7). The iPS cells grow indefinitely in laboratory cultures, and their developmental potential appears similar to that of ESCs: they can be made to differentiate into the body's vast array of different cell types.

Creation of iPS cells is exciting because they may someday replace ESCs in regenerative medicine. If this happens, the moral dilemma of using human embryos to obtain pluripotent cells will be solved. Initially, iPS cells had safety issues. The procedure to create iPS cells used viruses with health risks and a gene known to cause cancer. Dramatic progress in solving these problems came in 2009.[4] Mouse and human iPS cells were produced without using viruses, and a way was found to remove the added genes from the cells after genetic reprogramming was completed (Pera 2009).[5] Also, German and South Korean researchers created iPS cells by introducing a single, non-cancer-causing gene into human adult stem cells from nervous system tissue (Kim et al. 2009). However, work in 2011 uncovered several genetic abnormalities in iPS cells that currently preclude their safe use clinically (Pera 2011).

In view of iPS cells' pluripotency, is continued research on human ESCs needed? In addition to the safety issues with iPS cells, there are at least three reasons why research on human ESCs should be continued:

1. The enormous developmental potential of ESCs and the prospective contributions of ESCs to regenerative medicine are still "gold standards" against which all other types of pluripotent cells will be compared for years to come. Also, some biochemical indicators suggest that iPS cells do not equal ESCs in developmental potential and, therefore, in clinical value.
2. Human embryos and the stem cells derived from them will be invaluable for solving future problems in developmental biology, reproductive biology, and medicine; therefore, it is important to learn more about ESC biology and to continue examining the ethical issues surrounding their use in research and medicine.
3. As we see later, iPS cells bring up significant ethical issues of their own.

Therapeutic Cloning: What Is It and Why Do It?

Although many mammals have been cloned, humans have not yet been cloned. But this could change any day. Groups of ESC researchers around the world work toward cloning humans to the blastocyst stage in order to obtain ESCs for medical purposes. It is important to consider what cloning is and how the two types of cloning, *therapeutic cloning (embryo cloning)*

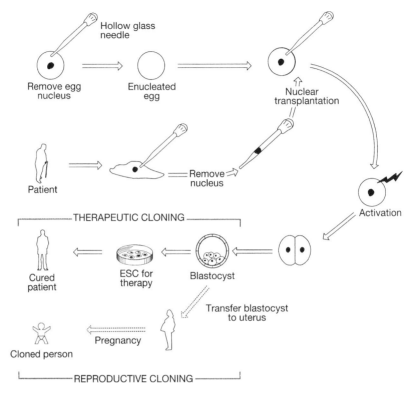

Fig. 2.8. Therapeutic and reproductive cloning. Therapeutic cloning of a patient to the blastocyst stage would allow a line of immunologically compatible ESCs to be obtained for repairing diseased or injured tissue. The nucleus would be removed from a woman's egg in the laboratory and replaced with the nucleus from a body (somatic) cell of the patient. The procedure is called somatic cell nuclear transfer. The egg receiving the somatic cell nucleus behaves like it has been fertilized by undergoing rounds of cell division that produce an embryo. At the blastocyst stage, the inner cell mass would be used to create an ESC line immunologically compatible with the patient. Alternatively, the blastocyst could be transferred to a woman's uterus for reproductive cloning, a procedure outlawed in several countries and US states.

and *reproductive cloning,* differ, remembering that for humans the descriptions are still hypothetical.

The aim of therapeutic cloning is to produce a blastocyst embryo genetically identical to a patient in need of ESCs for treating a disease or injury. By contrast, the objective of reproductive cloning is a birthed individual genetically identical to the individual being cloned.[6] In therapeutic cloning, cloned embryos grown in the laboratory would be used to obtain ESCs for therapy. The cloned embryos would never be transferred to a woman's womb. On the other hand, in reproductive cloning, cloned blastocysts would be transferred to a woman's uterus to produce a pregnancy.

Both types of cloning would employ a procedure called *somatic cell nuclear transfer (SCNT),* in which the nucleus from a donor cell is transferred into an egg cell that has had its own nucleus removed (fig. 2.8 and chapter 7).

Embryo cloning for ESC production has been successful in mice and monkeys, and it is probably just a matter of time before it is accomplished in humans.[7] But there are legal and technological hurdles to overcome before therapeutic cloning in humans becomes routine. For example, several countries and states in the United States prohibit human therapeutic cloning.

The advantage of obtaining ESCs from a patient's blastocyst clone is that the ESCs are perfectly compatible with the patient's immune system. The cells will not be rejected by the patient, as would be cells derived from embryos unrelated to the patient. If immunologically compatible ESCs can repair injured or diseased organs, there would be no need to suppress the patient's immune system with drugs as is now required with organ transplants. Patients needing bone marrow transplants after radiation treatment or chemotherapy for cancer or suffering from Parkinson's disease, type 1 diabetes, spinal cord injuries, or brain or heart injury due to strokes and heart attacks would all benefit from therapeutic cloning.[8]

Stem Cells and Embryos: Human Values

The moral status of human embryos is the most obvious, but not the only, values issue brought up by stem cell research. Other ethical questions raised by stem cell technologies include how to justly distribute their benefits, whether we should use ESCs to greatly increase human life span, and whether ESCs ought to be used to create genetically engineered human beings or alter the genetic diversity of human populations. Discussions of the last three questions are in chapters 4 (*eugenics* and genetic diversity), 5 (human genetic enhancement), and 7 (age retardation). Here we consider the moral status of human embryos and distributive justice issues.

Definitions for Informed Opinions

Good decision making requires accurate information, and reliable information requires a vocabulary appropriate for the problem at hand. One hindrance to informed discussion about the moral status of human embryos is ambiguous use of the terms *embryo, fetus, human life, human being,* and *person.* Without clear definitions we often talk past each other rather than with each other. The following definitions allow us to examine the ethical issues with a common vocabulary.

Embryo and *fetus.* Physicians and developmental biologists distinguish between human *embryos* and *fetuses* on the basis of developmental stage. Embryonic development begins with the fertilized egg (zygote) and continues for eight weeks. Fetal development includes the period from eight weeks until birth. The significance of the eight-week point is that by then organ systems have formed, though they still must grow and mature before becoming fully functional.

Human life. All living human cells manifest "human life." Human cells growing in a laboratory dish, cultured tissue fragments, organs, embryos, and fetuses are all examples of human life. Even eggs and sperm, whether within an individual or not, are forms of human life. So if a cell, or group of cells, has a human genome and is alive, it is human life, as distinguished from petunia or chipmunk life. Many types of human cells are grown in the laboratory, and thousands of cell biologists around the world study human life by researching such cells.

Human being. All forms of human life do not qualify as human beings. A human being is an organism belonging to the species *Homo sapiens.*[9] An organism is a cell or group of cells with interdependent parts that cooperate to perform vital functions and which itself is not simply part of some larger living entity. Amoebae, sponges, jellyfish, mushrooms, fertilized human eggs, and human embryos and fetuses are all organisms. Human embryos, whether inside the womb or in a laboratory culture dish, are therefore human beings. So too are all living, postnatal *Homo sapiens,* including those who are asleep, otherwise unconscious, comatose (with or without life support machines or brain waves), or suffering from advanced Alzheimer's disease. By contrast, a severed toe is not a human being since it was part of a larger living entity.

Person. All human beings may not qualify as persons. A person is a human being who possesses a special dignity conferred upon her or him by other human beings. With rare exceptions, the right to continued existence is one aspect of personhood. There is no consensus on criteria for personhood. For some, the fertilized egg and a brain-dead individual on total life support are persons, whereas, for others, one may need to be born or to have brain waves to qualify as a person. Certain biotechnologies raise the question of when during human development personhood should be assigned. We will examine different approaches and criteria for answering this question.

Human Embryo Research Raises Ethical Issues

Are human blastocysts and earlier-stage embryos persons? Three biotechnologies force this question: (1) ESC technology, (2) therapeutic cloning, and (3) *pre-implantation genetic diagnosis (PGD)*. All three require the manipulation and ultimate destruction or discarding of very early human embryos growing in glass dishes. Local, state, and national politicians, spokespersons for various religious groups, bioethicists, and others speak out about the moral status of human embryos. These voices influence the direction of biomedical research through the politics of its funding.

Embryonic Stem Cells: Personhood and the Human Embryo

Emotion and reason both have a place in making moral decisions. Strong feelings on issues less weighty than personhood, such as whether professional athletes should be allowed to strike or whether handguns should be registered, range widely, and reason does not guarantee agreement. Scientists often disagree on how to interpret the same experimental data, philosophers disagree on why an act is right or wrong, and voters disagree on what makes for good candidates. Can we find any common ground in the human embryo debate? I suggest that we start with knowledge about something that we all have in common—our development from a single cell. Knowing about stages of development that we all go through before birth can help us to understand why there are so many different views on when personhood begins and to answer this important question for ourselves.

Biological signposts during human development. Development of a fertilized egg into a newborn taking its first breath is a gradual process. *Conception* occurs when a sperm fuses with an egg cell and introduces its DNA-laden nucleus into the egg's cytoplasm. *Fertilization* itself occurs about twenty-four hours later when the sperm and egg nuclei fuse, an event called *syngamy,* which results in co-mingling of the paternal and maternal sets of chromosomes and creation of a *zygote,* the fertilized egg. Next, the DNA in the zygote nucleus replicates, and the cell divides to produce the 2-cell stage. Cell division continues through several cleavage stages until the embryo produces a solid ball of about 32 cells, the *morula.* By four to five days after conception, the embryo contains 100 to 150 cells in the form of a hollow ball, the blastocyst (fig. 2.9).

Embryo implantation into the wall of the uterus occurs at the blastocyst stage six to seven days after conception. The inner cell mass of the blastocyst eventually forms the fetus, while the outer wall of the blastocyst forms the fetal portion of the placenta. But 40 to 80 percent of all human blastocysts created by natural, internal fertilization never successfully implant and are simply sloughed out of the body during a woman's next menstruation (Nobel 2010).

By fourteen days, the inner cell mass is a flat disc containing thousands of cells. A ridge of cells forms along the length of the disc, and at one end the cells migrate from the surface of the embryo to its interior in a process called gastrulation. In mammals, gastrulation marks the *primitive streak* stage of development, during which the embryo acquires three distinct layers of cells. The outermost layer is the *ectoderm,* committed to developing into skin, the nervous system, teeth, and parts of the eye. The middle layer is the *mesoderm,* committed to forming skeletal and heart muscle, bones, and the circulatory, reproductive, and urinary systems. The inner layer is the *endoderm.* It will form the gastrointestinal tract, liver, pancreas, and

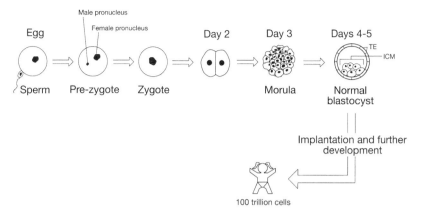

Fig. 2.9. Stages of early embryonic development in humans and other mammals. The sperm cell nucleus containing the male's genetic material enters the egg and fuses with the egg nucleus to form the nucleus of the zygote, a fertilized egg. DNA in the zygote nucleus replicates, and the zygote divides to produce a two-cell-stage embryo. DNA replication and cell division continue until the blastocyst-stage embryo forms by about four days after fertilization. In successful pregnancies, the blastocyst implants itself into the uterine wall and development continues. After implantation, the trophectoderm (TE) gives rise to the fetal component of the placenta, and the inner cell mass (ICM) gives rise to the fetus.

lungs. At the primitive streak stage of development, the embryo loses its capacity to split and form identical twins (or triplets) or to fuse with a fraternal twin embryo producing a single embryo.[10]

Cells destined to form the central nervous system are first identifiable at the primitive streak stage even though identifiable nerve cells are not yet present. On day 18, tissue rudiments of the future central nervous system appear as the neural plate, a layer of thickened cells on the back (dorsal) side of the embryo. At 22 days after fertilization, an embryonic, pipe-shaped heart begins beating. From here on, dramatic and easily observed changes occur rapidly (table 2.1). Between days 22 and 66, the embryo becomes recognizably human, forming limbs, fingers, toes, genitalia, and facial features. The central nervous system, including the cerebral cortex, the part of the brain required for reason, also begins to form during this period. Sensory nerve connections to the brain for pain perception are

Table 2.1. Biological signposts during human development

Development time*	Stage	Remarks**
0 hours	Conception	Sperm penetrates the egg cell membrane; sperm nucleus with paternal chromosomes enters the egg cytoplasm.
24 hours	Zygote	Sperm nucleus and egg nucleus fuse, marking fertilization or *syngamy*; maternal and paternal chromosomes intermingle; the cell is *totipotent*, i.e., it can give rise to an entire individual plus placental tissues.
30–48 hours	Cleavage	Zygote divides to produce 2 cells, 4 cells, 8 cells, etc. Cells are called *blastomeres* and are totipotent to about 8-cell stage, i.e., each blastomere can produce an entire individual if separated from the others.
48–72 hours	Morula	Fewer than 100 cells; individual cells are no longer totipotent but most are *pluripotent*, i.e., able to produce all tissues in a fully formed individual.
4–7 days	Blastocyst	Embryo one-fifth the size of a pinhead; normal time of implantation in uterine wall; produced in IVF clinics; cells of the inner cell mass (ICM) of surplus IVF blastocysts are potential sources of new embryonic stem cell lines; ICM contains 150 to a few hundred cells that can produce a fetus, whereas cells surrounding ICM produce placental tissue; all ICM cells are pluripotent.
14 days	Primitive streak	Embryo the size of a poppy seed; rudiments of three major classes of embryonic tissue (*ectoderm, mesoderm, endoderm*) formed; embryo can no longer split to form twins, cells are no longer pluripotent; *individuation*.

Development time*	Stage	Remarks**
18 days	Neurula	Rudiments of central nervous system forming.
22 days		Embryo is size of a sesame seed; heart begins beating; from about this stage onward, differentiated tissues contain *multipotent* stem cells, i.e., *adult stem cells* able to produce different cell types found in the tissue in which they reside.
24 days	Pharangyla	Appearance of this stage is similar in all vertebrates; rudimentary tail and gill slits are present; earliest cells of the future brain and spinal cord are forming.
31–32 days		Embryo about the size of a lentil; first nerve cells of the future cerebral cortex appear; arm and leg buds appear.
35 days		Embryo about the size of a blueberry; olfactory nerve forms connections in the brain; this is the first sensory nerve to form synapse with brain cells.
40–44 days	"Quickening"	Embryo about the size of a kidney bean; first muscular movements; eyelids, fingers, and nose appear.
45 days		First brain waves, but not the type marking consciousness.
56 days	Embryo becomes a fetus	Fetus about the size of a kumquat; organ systems are formed by eight weeks, but they are not functionally mature.
60 days		Gender is recognizable from the genitalia.
66 days		Fetus about the size of a fig; human-like facial features are present.

Continued on the next page

Table 2.1. *Continued*

Development time*	Stage	Remarks**
90 days	End 1st trimester	Fetus about the size of an apple; bones of the cranial vault and face are developing.
130 days		Fetus as long as a banana; nerves for sensing pain are present.
180 days	End 2nd trimester	Fetus as large as a cauliflower head; viability possible outside the uterus.
200 days		Fetus weighs as much as four navel oranges; first brain waves characteristic of consciousness.
270 days	Neonate	Birth; first breath.
2–4 years	Postnatal	Sense of a life narrative with past, present, and future; "othermindedness" (empathy, deception possible).
10 years		Cerebral cortex (the "thinking brain") fully functional.

*Times in this chart begin with fusion of egg and sperm; physicians commonly mark development beginning two weeks earlier, from the time of the woman's last menstruation.
**Size comparisons are taken from the Pregnancy Development Calendar at http://www.babycenter.com/pregnancy-calendar (accessed December 14, 2011).

complete by 130 days; however, pain probably cannot be experienced until after the second trimester at about 200 days, when brain waves characteristic of human wakefulness first appear.

At six months, *in utero* connections between nerve cells and the cerebral cortex, the part of the brain responsible for human consciousness, form at an accelerated rate. At birth, a few seconds after the umbilical cord is tied off, the infant takes its first breath. Awareness of a past, present, and future (a life narrative) emerges at about two years after birth, ability to empathize and deceive ("other mindedness") at about four years, and a fully functional cerebral cortex at about ten years of age. Table 2.1 summarizes these and other developmental events.

Approaches to the personhood question. Three views about personhood underpin decisions about the moral status of human embryos: the single-criterion view, the pluralistic view, and the potentiality view. The single-criterion approach identifies a single, decisive factor for granting personhood. The criterion may be conception, a beating heart, viability outside the womb, or some other characteristic of the developing embryo or fetus. When that characteristic first becomes present, personhood is established. The pluralistic approach views personhood as emerging gradually during embryonic, fetal, or postnatal development. This view lends itself to assigning increasing levels of moral status to embryos and fetuses as they accumulate more and more of the set of traits that marks full personhood.[11] Finally, the potentiality approach assigns moral status to the embryo based on what it can become rather than what it actually is.

(1) Single-criterion views on personhood. Single-criterion views on personhood come from both religious and secular realms. A frequent religion-inspired criterion for personhood is the presence of a soul; secular views usually focus on some biological attribute of the developing individual. Let's look at some examples of each approach.

Prior to the thirteenth century and the influence of Thomas Aquinas (1225–1274), Christianity's view on the beginning of personhood is unclear. But with Aquinas, who adopted Aristotle's view on when the rational soul enters a human being, the Christian church maintained that the human embryo becomes a person on the fortieth day after conception.[12] Christianity replaced Aristotle's rational soul with a God-given immortal soul.

Now Roman Catholicism marks the beginning of personhood with conception. How did that change occur? In the latter part of the nineteenth century, German and Belgian biologists showed that both plants and animals begin life as fertilized eggs. Responding belatedly to this biological information, the Roman Catholic Church declared in 1917 that human embryos must be treated as persons from conception onward. The church does not claim to know precisely when ensoulment occurs, but it chooses to give full moral status to the conceptus in order to be on the safe side. In August 2000 the Pontifical Academy for Life reiterated the Catholic Church's position that personhood begins at conception and condemned the use of human ESCs because human embryos are destroyed:[13]

> The Church has always taught and continues to teach that the result of human procreation, from the first moment of its existence, must be guaranteed that unconditional respect which is morally due to the human being in his or her totality and unity in body and spirit: The human being is to be respected and treated as a person from the moment of conception; and therefore from that same moment his rights as a person must be recognized, among which in the first place is the inviolable right of every innocent human being to life. (Pontifical Academy for Life 2000)

Joining the Roman Catholic Church in giving the laboratory-generated conceptus personhood status and opposing ESC research and use are the Holy Synod of Bishops of the Orthodox Church in America and the Southern Baptist Convention. Unclear is whether these institutions considered the twenty-four-hour lag after conception and before fertilization while forming positions on personhood.

Other Christian denominations take stands in the ESC debate without formal declarations on the beginning of personhood. The 2001 General Assembly of the Presbyterian Church (USA), a group of elected representatives from the denomination, endorsed the use of ESCs for research and therapy so long as they are derived from "embryos that do not have a chance of growing into personhood because the woman has decided to discontinue further (IVF) treatments and they are not available for donation to another woman for personal or medical reasons, or because a donor is not available" (Presbyterian Church 2001). The 2004 General Conference of the United Methodist Church adopted a similar position, resolving that the conference go on record in support of those persons who wish to enhance medical research by donating their early embryos remaining after *in vitro* fertilization (IVF) procedures have ended . . . for ESC research provided that

1. these early embryos are no longer required for procreation by those donating them and would simply be discarded;
2. those donating early embryos have given their prior informed consent to their use in stem cell research;
3. the embryos were not deliberately created for research purposes;
4. the embryos were not obtained by sale or purchase.

(The people of the United Methodist Church 2001)

The Islamic tradition also uses ensoulment as the single criterion for conferring full moral status to developing human beings; Islamic jurists place the event either at 40 or 120 days after conception. In either case, Islam considers our ability to use ESC technology to relieve suffering as God-given and therefore justifiably good.

In Judaism, the moral status of the embryo is elevated after 40 days of development, but not to the level of ensoulment, the time for which is considered unknowable except by God. Beginning on day 41, the embryo is assigned a moral value equivalent to that of a major body part of the mother. The Talmudic basis for this and the Jewish position on ESC research was articulated in 2001 by Rabbi Elliot Dorff of the University of Judaism in Los Angeles. He wrote that God requires that "we seek to preserve human life and health . . . and have a duty to seek to develop new cures for human diseases" (Dorff 2001). Referring to surplus embryos resulting from IVF procedures, Rabbi Dorff contends that "when a couple agrees to donate such embryos for purposes of medical research, our respect for such pre-embryos outside the womb should certainly be superseded by our duty to seek to cure diseases" (2001).

Departing from Vatican authority, Salesian priest, theologian, and philosopher Norman M. Ford (1988) marks the beginning of personhood at the primitive streak stage. He argues that here, at day 14, the embryo acquires full moral status because it is on a trajectory leading to the development of one particular individual.

Single criteria for personhood status in human embryos also come from the secular realm. A US government panel on embryo research advised adherence to a 14-day rule for the latest stage at which human embryos can be used in medical research because 14 days after conception is when some cells (ectoderm, table 2.1) in the embryo are committed to develop into parts of the central nervous system, including the brain (National Institutes of Health 1994, 67). At day 14 a pain-sensitive, rational brain is still many months away, so the panel played far on the safe side with its 14-day rule.[14]

Viability of the fetus is a criterion for personhood with the status of federal law in the United States since the 1973 Supreme Court decision in *Roe v. Wade*. The Court's decision defines the first two trimesters as a period during which women cannot be denied an *elective abortion*. Six months into a pregnancy is still about the earliest that modern neonatal units can sustain fetal life outside the womb.[15]

Neuroscientist Michael Gazzaniga, an expert in human brain function, also embraces viability as a criterion for acquisition of personhood. He writes that "in judging the fetus 'one of us,' and granting it the moral and legal rights of a human being, I put the age . . . at 23 weeks, when life is sustainable and the fetus could, with a little help from a neonatal unit, survive and develop into a thinking human being with a normal brain" (Gazzaniga 2005, B10–B12). Gazzaniga's emphasis on brain function reflects the use of human consciousness as the defining mark of personhood.

(2) Pluralistic views of personhood. Before declaring conception as the single criterion for personhood, the Roman Catholic Church had a pluralistic view of personhood as reflected in sanctions against abortion. For example, some seventh-century books of penance require a greater level of penance for aborting a "formed" fetus (forty days old) than an "unformed" fetus (earlier than forty days). From 1591 to 1869, canon law dictated ex-

communication for a woman aborting a "formed" fetus, but not for an "unformed" fetus. These rules of penance suggest a belief that fetuses acquire greater levels of moral status as they progress developmentally. In fact, in the late sixteenth century, Catholic theologian Thomas Sanchez argued that the developing embryo goes through graded levels of moral status (in D. Jones 2001). Only in 1869 did the Catholic Church remove the distinction between "formed" and "unformed" fetuses and prescribe excommunication for aborting a fetus at any stage of development.

A strongly voiced primary criterion for personhood can mask underlying pluralistic thinking. For example, the 1999 Southern Baptist Convention passed a resolution declaring that "protectable human life begins at fertilization," presumably reflecting the belief that an immortal, God-given soul is present at that time. But in the same document that declares opposition to the use of human blastocysts to obtain ESCs and describes human embryos as "the most vulnerable members of the human community," the convention reminds its denomination that abortion can be sanctioned "to save the physical life of the mother" (Southern Baptist Convention 1999). The latter position reveals pluralistic thinking since it implies that the moral status of the mother is greater than that of the fetus. The document does not detail why the mother is given greater consideration, but one might assume that it is because she has accumulated more hallmarks of personhood, such as rationality and a life narrative, than has the fetus.

(3) Potential personhood. Assigning personhood to an early embryo does not require belief that it has a God-given soul. One can simply declare that an embryo has the potential to develop into a person and that this biological potential alone entitles it to the moral status of personhood. Some bioethicists speak of "latent potential" versus "active potential," giving greater moral status to the latter. For example, an unfertilized egg and a nearby sperm cell have only latent potential to give rise to a person, compared to the active potential of a fertilized egg to do the same.

The problem with distinguishing between latent and active potentials is that the line between them is necessarily arbitrary. For example, one could say that the potential of an embryo frozen at an IVF clinic is "latent"; whereas, that of one transferred to a uterus is "active." Or once inside the uterus, a preimplantation blastocyst is latent in its potential for personhood, and an implanted blastocyst is active.

For bioethicist John Harris (1985), assigning personhood to a cell or group of cells on the basis of its potential to develop into a person leads to absurdities. For example, all human sperm and eggs are potential progenitors of persons. Should we strive to help every egg and sperm cell realize its potential? Put in the hands of skilled cloning technicians, potential persons also reside inside the thousands of *somatic cells* that we kill and rinse down the drain after bathing or brushing our teeth. Harris argues for conferring personhood to entities based on what they actually are, not on where they are located or what they might become.

Balancing benefit and harmfulness in the embryo debate. In 2001, resolutions presented to the General Synod of the United Church of Christ and to the General Assembly of the Presbyterian Church (USA) called upon the representative bodies of both denominations to urge support for expanded ESC research by the US government. The resolutions cited the good that ESC research could bring to humankind through innovative and lifesaving diagnoses and treatments and cures for diseases and injuries. The resolutions reflect a judgment that the harm ESC research causes to surplus embryos that have no chance of developing further is outweighed by the potential of the research to relieve human suffering. The Orthodox Jewish tradition also favors ESC research for its potential to heal, giving priority to helping the sick and injured over protecting unconscious embryos grown outside the womb.

Christian theologians Ted Peters (Pacific Lutheran Theological Seminary) and Gaymon Bennett (Graduate Theological Union, Berkeley) (2003) urge that the human embryo debate be reframed in terms of beneficence, away from its present fixation on the moral status of unconscious clumps of cells. They argue that the Christian tradition focuses on healing and helping persons become who they were intended to be rather than on examining our embryological origins to discover our moral selfhood.

Therapeutic cloning, Immanuel Kant, and slippery slopes. Nineteenth-century German philosopher Immanuel Kant has had a tremendous influence on how we treat each other today. One of the greatest moral philosophers of the modern era, Kant formulated a rule for judging the morality of actions between humans: "Act as to treat humanity, whether in thine own person or in that of any other, in every case as an end withal, never as means only" ([1785] 2000, 297). This is called Kant's categorical imperative. What it means is that we ought to respect the autonomy of every person to make of her or his life what she or he decides and that we must not treat others merely as means to benefit ourselves.

What do Kant's ideas have to do with using therapeutic cloning to obtain ESCs? For those who believe that blastocysts are persons, creating a blastocyst *solely* for research or therapy, no matter how great the benefit and no matter how many people are helped, violates Kant's dictum. On the other hand, persons favoring therapeutic cloning can argue that it would not violate Kant's categorical imperative because the embryo does not have conscious personal interests and therefore has no basis for Kantian autonomy. A reply to this argument might point out that a newborn has no better idea about its self-interest than does a blastocyst, yet we do not use newborns solely for research or therapy.

Some persons who support ESC research oppose therapeutic cloning because they believe it is a slippery slope leading to reproductive cloning, the subject of chapter 7. Presuming for the moment that reproductive cloning is a bad thing (a contentious presumption examined in chapter 7), the notion that creating cloned blastocysts for therapy should be banned because it might lead to cloned babies is like saying we should stop teaching chemistry in high school because some students may use the knowledge to synthesize illicit drugs. Slopes can always be terraced so that we don't slip. We give our will and intellect too little credit if we believe that embarking on a program to develop therapeutic cloning to help victims of disease and traumatic injury will lead inevitably to reproductive cloning. Will and intellect notwithstanding, at this writing there is pending federal legislation in the United States to prohibit both therapeutic and reproductive cloning.

National and state policies on human embryo and ESC research. A population of ESCs derived from a single blastocyst is called an *embryonic stem cell line.* Since 1998, dozens of ESC lines have been established in laboratories at universities and in private research institutes, mainly in the United Kingdom, Israel, China, Belgium, South Korea, and in California, Connecticut, Wisconsin, Illinois, Maryland, and New Jersey.

Laws regulating ESC research vary widely between countries and between states in the United States. They are subject to the influences of religious traditions, national history, and political and economic forces.

Israel is among nations with the least restrictive policies regulating human embryo research. It allows the creation of new ESC lines from surplus embryos and also from embryos created explicitly for research. Israel also allows therapeutic cloning. These policies are consistent with the Jewish religious view of the moral status of early human embryos and reflect Israel's Jewish majority.

Other nations with policies similar to Israel's include the United Kingdom, Belgium, India, and South Korea. In China, both therapeutic cloning and the transfer of human cell nuclei into nonhuman animal eggs are allowed. The latter practice helps resolve ethical problems associated with collecting and using human eggs for research. Correlations between religious traditions and national polices are not clear in these nations, but the prospect for future economic payoffs from human ESC technologies is probably an important factor in their policy making.

European Union (EU) science ministers agreed in 2006 to allow funding from the European Union's science budget for member nation research on preexisting ESC lines, but not all EU member countries allow such research. Ireland, Italy, and Poland, all nations with Roman Catholic majorities, prohibit research with human embryos, including research on ESC lines already in existence. Austria and non-EU member Norway have similar policies. In Germany, creating new ESC lines is illegal, but research on existing ESCs imported into the country is permitted. National history and philosophy are especially important in shaping Germany's policy on human embryo research. Still recovering from the horrific legacy of the

Holocaust, and influenced by the philosophy of its native son, Immanuel Kant, Germany strives to avoid even the appearance of insensitivity when it comes to using human beings in medical research.

In no country is funding for ESC research more subject to the winds of politics than it is in the United States. On August 9, 2001, President George W. Bush declared that no federal funds could be used for research on ESCs created after that date.[16] Twice the US Congress passed legislation to end the 2001 presidential restrictions on ESC research, and twice the president vetoed the legislation. On March 9, 2009, President Barack Obama revoked the Bush policy on ESC research and allowed federal grant researchers to work on ESC lines created since the 2001 prohibition so long as creation of the cell lines follows federal ethical guidelines. This is a boon to ESC researchers in public institutions and hospitals in the United States because many ESC lines were created by privately funded research during the eight years of the Bush restrictions.

Federal funds cannot be used to actually create new ESC lines in the United States. Doing so would violate the Dickey-Wicker Amendment passed by the (Newton) Gingrich Congress near the end of President William (Bill) Clinton's first term in office. The amendment prohibits use of federal monies for the creation or destruction of human embryos for research purposes, and the legislation has been renewed every year since 1995. While Dickey-Wicker restricts the use of federal dollars for ESC research, no US federal law prohibits using state or private monies to create new ESC lines or to pursue therapeutic cloning.

Revised National Institutes of Health (NIH) guidelines for federally funded human ESC research were instituted on July 7, 2009 (National Institutes of Health 2009). The new guidelines allow researchers to apply for NIH grants to work on human ESCs that were derived from surplus embryos produced by IVF for reproductive purposes. However, the NIH does not support research that introduces human ESCs into nonhuman embryos or that uses human ESCs created by SCNT (i.e., cloning, fig. 2.8), by parthenogenesis (embryo creation without the use of sperm), or from embryos created solely for research. In June 2010, some prolife groups, including two adult stem cell researchers, sued the NIH, claiming that the new NIH guidelines violate the Dickey-Wicker Amendment because NIH funding of ESC research is preconditioned on the destruction of human embryos. In August 2010, US District Court judge Royce Lamberth placed an injunction on federally funded ESC research while he considered the legality of the Obama administration's policy on ESC research. On April 29, 2011, a Federal Court of Appeals reversed the lower court's injunction, allowing NIH funding of ESC research to continue while Lamberth continued his deliberations. Finally, on July 27, 2011, Lamberth threw out the suit, ending the dispute and confirming the legality of the Obama administration's funding of ESC research (Wadman 2011).[17] Still, the past decade of political and legal controversy surrounding ESC research severely damaged the field in the United States by forcing many scientists away from the field and discouraging an untold number of young researchers from entering it.

Several states have laws regulating research on human embryos and fetuses. Some states' laws are more permissive than the Dickey-Wicker Amendment while others are more restrictive.[18] For example, in California, Connecticut, Illinois, Maryland, Massachusetts, Missouri, New Jersey, Rhode Island, and Wisconsin, the derivation of human ESCs from surplus IVF embryos is allowed. But Louisiana and South Dakota prohibit research on any cells derived from embryos outside a woman's body; the Louisiana law specifically prohibits research on surplus embryos created by IVF. So, in these two states, researchers cannot work with human ESCs regardless of when or where the cells were derived. In November 2011, voters in Mississippi rejected a proposed amendment to their state constitution that would have declared personhood to begin at the moment of fertilization, a definition of "person" that would have prohibited derivation of ESCs from surplus IVF embryos and abortion under any circumstances. It also would have called into question the legality of contraception methods like some IUDs that can prevent embryos from implanting in the uterus.

Adult Stem Cells: Ethical Issues

Using non-fetal, adult stem cells for research and therapy avoids the controversy of personhood. The main ethical issue for adult stem cell research and its possible future medical applications is how to fairly distribute the benefits of the technology.

Embryonic Germ Cells: Ethical Issues

Recall that human EGCs are derived from the primordial germ cells of late embryos or early fetuses aborted five to ten weeks after conception. Research using EGCs poses major ethical questions about the source of tissue for the research, particularly whether fetuses aborted electively ought to be used for such research. Fortunately, EGCs appear to have no clinical value beyond that of ESCs, so the ethical problems are largely hypothetical. Even so, basic research on EGCs can provide valuable information about egg and sperm cell formation, including insights into causes of infertility. But basic research in these areas can progress adequately using nonhuman animal models and, in rare instances, tissue from fetuses aborted for therapeutic reasons where parents have consented for the tissues to be used in this way.

Can Technology Solve the Moral Dilemma of Having to
Destroy Embryos to Obtain Pluripotent Stem Cells?

Scientists and ethicists have developed or proposed several ways to obtain pluripotent, human cells without destroying healthy, human embryos (President's Council on Bioethics Washington 2005). These include using cells from abnormal IVF embryos that could never develop into a fetus, using single cells from cleavage-stage IVF embryos without preventing the rest of the embryo from completing development, and using embryos created by transplanting human DNA into rabbit or other nonhuman eggs. Although all of these are technologically feasible, not one of them completely solves the personhood problem plaguing ESC research.

By far the most promising alternatives to ESCs are the iPS cells described earlier, normal body cells induced to become pluripotent by the addition of one or more genes. However, even iPS cells are not free from major ethical problems.

One ethical issue arising from iPS cell technology is that quasi "cloning" is possible. Both eggs and sperm can be derived from male iPS cells. Therefore, a man could have a child, with the aid of a surrogate mother, produced by the union of his own sperm and eggs. Similarly, a gay couple could have their own child, one member of the couple supplying the sperm and the other supplying the egg. But these unusual reproductive options would be open only to male donors, not women or lesbian couples whose cells lack the Y chromosome needed for sperm formation.

Another ethical issue for iPS cell research is shared with ESCs. Since iPS cells are pluripotent, they theoretically can be mixed with other pluripotent cells in blastocyst embryos to create *chimeras,* organisms containing cells from two different individuals. Even chimeras containing a mixture of human and animal cells would be possible.

Japan's Science Ministry has responded to these technological possibilities by forbidding (1) the transfer of embryos with iPS cells into human or animal wombs, (2) the creation of any individual using iPS cells, (3) transferring iPS cells into an embryo or fetus, and (4) using iPS cells to produce eggs or sperm. At this writing (early 2012), no regulations govern iPS cell use in the United States.

Distributive Justice and Research Priorities

Even if the problem of the moral status of human embryos were to be resolved, ESC research would still raise other important ethical issues. Among these are *distributive justice* and research priorities.[19] Both are important issues for each of the biotechnologies examined in this book.

Distributive justice refers to the fair allocation of the benefits and burdens of economic activity. All systems of distributive justice must address the same difficult problems, which include deciding who should pay the cost of research and development for new technologies, who will be first to reap their benefits, how long it will take for the benefits to become widely accessible, and whether some persons will be excluded from the benefits. Related questions are (1) whether the technologies will be so expensive that only wealthy persons can afford to take advantage of them and (2) what the relative roles of government and private industry should be in making benefits more accessible.

A strict egalitarian view of distributive justice requires that all members of society receive the same degree of benefits. The late moral philosopher John Rawls devised a different approach called the "Difference Principle" (1993, 5–6). Rawls allows for an unequal distribution of the products of economic activity so long as their allocation improves the long-term well-being of the least favored. In other words, a just distribution of material benefits requires that social and economic inequalities be arranged so they are to everybody's advantage. For example, highly educated physicians who oversee the health of the community may justly earn salaries higher than persons with less education or who provide nonessential services. Applying the Difference Principle to ESC therapy would require that if only a few received ESC therapies, others must reap tangible benefits from the few who were treated.

Other writers place little emphasis on how material goods are distributed, focusing instead on the overall welfare of each individual. Libertarian-minded writers such as Robert Nozick (1932–2002) generally consider personal liberty and self-ownership to be morally more important than the distribution of either material goods or welfare. The libertarian view would probably place minimal importance on making benefits from biotechnology like ESCs available to persons other than those who could afford the cost. Other views of justice are especially sensitive to the under-representation of women or minorities in the marketplace and the need to justly reward indirect contributions to the economy, such as rearing children. Persons with these views will probably wish to make the benefits of new biotechnologies available to all contributors to the welfare of society, regardless of income.

Related to the distributive justice issue is prioritization of public-sponsored research. Researchers' time and public monies for health-related research and programs are finite. With millions of children in desperate need of vaccinations, clean water, shelter, proper nutrition, and emotional care, what portion of our resources should be used to develop technologies aimed at helping relatively few people, many of whom are aged by present standards? How should the expense of developing new biotechnologies to ease human suffering and extend productive human lives be balanced against the urgency of both long-standing and new needs to battle causes of current human suffering like HIV, hunger, malaria, and a global outbreak of wheat rust that threatens world food supplies? This question has no absolute answer, but the problem deserves a thoughtful solution.

Conclusion

Human embryonic stem cells (ESCs), embryonic germ cells (EGCs), and induced pluripotent stem (iPS) cells are all pluripotent—able to produce all cell types in the body. Adult stem cells generally have less developmental potential than ESCs and are multipotent—normally producing cell types just in the tissue in which they occur. ESCs come from surplus, five- to seven-day-old blastocyst embryos produced at in vitro fertilization clinics. EGCs come from tissue in five- to ten-week-old embryos or fetuses. And iPS cells come from genetically reprogrammed skin or other body cells. Of all stem cell types currently investigated, considering both biological and ethical factors, ESCs and iPS cells have the greatest potential for future clinical applications. They produce tissues derived from all three embryonic germ layers (ectoderm, mesoderm, and endoderm). Results from mouse experiments show that ESCs are very likely to be useful in treating heart disease, stroke, Parkinson's disease, diabetes, and spinal cord injuries in humans. In addition to treatments for traumatic injuries, degenerative diseases, and age-related maladies, ESCs have the potential to revolutionize pharmaceutical research and the understanding of causes for congenital abnormalities. And iPS cells may prove to be just as useful as ESCs, but further work is needed to confirm this.

Embryos are destroyed in the process of obtaining ESCs, and disagreement over the moral status of human embryos makes ESC research controversial. Therapeutic cloning and preimplantation diagnosis (chapter 3) also raise the issue of moral status for human embryos. Religious and secular beliefs and cultural traditions manifest a wide range of views on when, during development, humans deserve the moral status of personhood. Positions range from treating the fertilized egg as a person to withholding the status of full personhood until birth. Global or national consensus on the embryo-personhood issue seems unlikely. Influences affecting policies regulating human embryo research and ESC use will likely continue to include the weighing of possible benefits against harms resulting from the research, national histories, electorates' views, economics, and national pride in scientific discovery. Several technological solutions to the ethical problems of using viable human embryos to obtain ESCs have been proposed, but most of these are also controversial. One exception may be the genetic reprogramming of somatic cells to a pluripotent state similar to that of ESCs. An informed public can provide the wisdom and insight needed in policy debates and decisions affecting the future of regenerative medicine.

Questions for Thought and Discussion

1. When do you believe a human being first acquires full moral status as a "person?" Why? What specific rights should your person have? How would your answers to these questions be viewed by various groups within our pluralistic global village?

2. Is the blastocyst-stage human embryo a person?

3. Should federal agencies like the National Institutes of Health and the National Science Foundation support embryonic stem cell (ESC) research, including the creation of new ESC lines from surplus embryos generated by in vitro fertilization?

4. Are there scientifically and morally credible alternatives to using human embryos to obtain pluripotent cells for treating disease and injuries? If so, are there good reasons to continue doing research with human ESCs?

5. Assuming that an ethically acceptable, but limited, source of pluripotent cells for human regenerative medicine is developed, how should the use of this technology be prioritized; that is, who should benefit first and most from the technology and in what situations should it be used? Can you imagine medical or nonmedical uses of the technology that would be unethical?

6. Should therapeutic human cloning be pursued for use in regenerative medicine?

Notes

1. The article notes the timing of the FDA announcement in the same month as President Barack Obama took office. It reports that Robert N. Klien, chairperson of the $3 billion California stem cell research program, believes that pressure on the FDA from the Bush administration delayed the trial's approval that had been sought since March 2008.

2. See http://www.news-medical.net/news/20091031/Geron-plans-to-advance-clinical-program-for-spinal-cord-injury.aspx (accessed July 23, 2010); and http://www.mercurynews.com/ci_15641176?source=most_viewed (accessed July 30, 2010).

3. Developing humans are called embryos through eight weeks of development and fetuses from then until birth.

4. The Wisconsin research group substituted the Lin28 gene for the cancer-causing c-myc gene and obtained human iPS cells from skin fibroblasts. The Japanese research group used a circular DNA molecule called a plasmid to transfer pluripotency-inducing genes into mouse cells, but the reprogramming efficiency was much lower than when viral vectors are used. A collaborative team from Canada and the United Kingdom succeeded in genetically reprogramming mouse and human skin cells with high efficiency without using viral or plasmid vectors, and the same group showed with mouse cells that the added genes needed to induce pluripotency can be removed from the cells without a trace after reprogramming is accomplished.

5. Pera's article is a short summary of the following two research articles by the Canada-UK team that report reprogramming skin cells into iPS cells without using viruses or plasmids and removing the added genes from the cells after reprogramming is complete: Woltjen et al. 2009; Kaji et al. 2009.

6. Cloning by somatic cell nuclear transfer involves use of an enucleated donor egg that still possesses its own mitochondria and mitochondrial DNA harboring about two dozen genes. Unless donor eggs come from the individuals being cloned, whether they are human or other animals, the mitochondrial DNA in the clone will be slightly different from the mitochondrial DNA in the cloned individual; whereas, the nuclear DNA in the clone and in the cloned individual

has been thought to be identical. Human mitochondrial DNA consists of about 16,500 base pairs, while human nuclear DNA has 3.1 billion base pairs. Therefore, the DNA of a human clone, created from an egg not coming from the cloned individual, would be about 99.99947 percent identical to the DNA of the cloned individual. In 2008 it was also reported that identical twins do not share exactly the same nuclear DNA (*New York Times,* March 11, "The Claim: Identical Twins Have Identical DNA," by Anahad O'Connor), a finding that makes it likely that there would also be slight differences between the nuclear DNA of a clone and of the cloned individual.

7. Earlier reports of therapeutic cloning turned out to be fraudulent. In February of 2004, South Korean stem cell researcher Woo Suk Hwang and his coworkers reported having created a blastocyst-stage human clone and then derived ESCs from it. In June of the following year the group reported having produced eleven ESC lines derived from persons with spinal cord injuries, a child with juvenile diabetes, and a boy with an immune disorder. The work was published in *Science* magazine, one of the world's premier science journals. The report created tremendous excitement in the biomedical community because of likely application of the technology in treating traumatic injuries and numerous degenerative diseases. Shockingly though, by January of 2006, an investigative committee at Dr. Hwang's university in Seoul concluded that both reports were based on fabricated claims and misrepresented data. As regrettable as this was for ESC research, the saga does illustrate the self-correcting nature of science, which depends upon honesty and trust within the scientific community.

8. One limitation of therapeutic cloning is its inability to treat genetic diseases without the additional technology of gene therapy (chapter 6). All of the patient's cells would carry the genetic defect.

9. Tooley (1983, 51) defines *human being* as a "member of the biological species, *Homo sapiens.*" I prefer to stipulate that a human being is an *organism* in order to distinguish "human being" from simple, cellular "human life."

10. An embryo formed by the fusion of two embryos is called a chimeric or mosaic embryo. Chimeric embryos develop into chimeras, individuals whose bodies contain two genetically distinct populations of cells. Fusion of a male and a female embryo results in a mixed-gender chimera.

11. Specific traits used to recognize full-fledged personhood vary among ethicists and religious thinkers. They may include brain activity distinctive to consciousness, living free of the mother's body, sense of a life narrative, and the ability to reason morally.

12. Aristotle proposed three souls for humans: a nutritive soul that guides basic activities essential for life, including food intake, growth, and reproduction; a sensitive soul, possessed by all animals, that permits perception of the environment and movement; and a rational soul responsible for our reason and ability to symbolize. According to Aristotle, all animals possess nutritive and sensitive souls, but only humans have a rational soul. Interestingly, Aristotle and Thomas Aquinas taught that ensoulment at forty days applied only to male embryos and that female embryos required ninety days of development before the inviolable event occurs.

13. The Pontifical Academy for Life, founded in 1994, is an institution within the Roman Catholic Church. It promotes the church's position that all human life is sacred and should be protected. It also acts as a think tank on bioethics and Catholic moral philosophy.

14. The first sign of brain waves indicative of human consciousness appears at about six months into development, and the first sensory nerve connections in the brain are olfactory and occur at five weeks (see table 2.1).

15. The onset of viability may vary by a week or two, depending upon the fetus and the neonatal care given to it, but there are biological constraints on how far back technology can push viability. For example, one limiting factor is the development of the lungs to a point at which they can exchange oxygen between the atmosphere and the bloodstream. It is highly unlikely that technology to sustain fetuses much younger than six months will be developed in the foreseeable future.

16. On August 9, 2001, President George Bush prohibited use of federal funds to support research that involves creation of new embryonic stem cell lines from human embryos. This meant that no money from granting agencies like the National Institutes of Health or the National Science Foundation could fund this type of research. The prohibition did not apply to private or state monies. The

president cited two reasons for restricting federally supported ESC research: (1) destroying blastocyst embryos to create ESCs is disrespectful of human life, and (2) enough ESC lines are already available for researchers to use. The prospect for new adult stem cell therapies was also cited to support restrictions on ESC research. An inconsistency in the 2001 ban on producing new human ESC lines is that it does not address the eventual discarding of hundreds of thousands of surplus embryos created at IVF clinics. Most funding for research on ESC lines created after the president's prohibition came from state coffers. California, Wisconsin, Connecticut, Illinois, Maryland, and New Jersey all allocated funds for ESC research that included creation of new ESC lines. The largest state commitment was in California, where, in 2004, 59 percent of voters approved a $3 billion, ten-year bond issue to fund ESC research, including therapeutic cloning. Issuance of grants from this money was held up in the courts, but in 2006 the state's Republican governor, Arnold Schwarzenegger, provided $150 million for the research directly from the state budget on the day after President Bush vetoed legislation that would have removed his earlier prohibition on the research. On the same day, Democratic governor Rod R. Blagojevich of Illinois provided $5 million for ESC research in his state.

17. More information about the stem-cell lawsuit is available at http://www .nature.com/news/specials/stemcellinjunction/index.html (accessed May 8, 2012).

18. The National Council of State Legislatures published a table in 2008 detailing how laws in thirty-three states regulate research on human embryos and fetuses. http://www.ncsl.org/default.aspx?tabid=14413 (accessed November 11, 2011).

19. A vast literature exists on distributive justice. Two especially accessible resources on the subject are Lamont and Favor 2009; Beauchamp and Childress 2001, 225–82.

Sources for Additional Information

Barry, Patrick. 2007. "Hold the Embryos: Genes Turn Skin into Stem Cells." *Science News* 172 (21): 323.

Dorff, Elliot N. 2001. "Embryonic Stem Cell Research: The Jewish Perspective." Responsum presented to the Committee on Jewish Law and Standards, December. Published in the *United Synagogue of Conservative Judaism Review* (Spring 2002). http://www.oca.org/QA.asp?ID=68&SID=3 (accessed January 30, 2008).

Ford, N. M., 1988. *When Did I Begin? Conception of the Human Individual in History, Philosophy, and Science.* Cambridge, UK: Cambridge University Press.

Garreau, Joel. 2005. *Radical Evolution.* New York: Doubleday.

Gazzaniga, M. S. 2005. "The Thoughtful Distinction Between Embryo and Human." *Chronicle of Higher Education,* April 8, B10–B12.

Holland, Suzanne, Karen Lebacqz, and Laurie Zoloth, eds. 2001. *The Human Embryonic Stem Cell Debate: Science, Ethics, and Public Policy.* Cambridge, MA: MIT Press.

Holy Synod of Bishops of the Orthodox Church in America. 2001. "Embryonic Stem Cell Research in the Perspective of Orthodox Christianity." October 17. http://www.orthodoxresearchinstitute.org/articles/ethics/oca_embryonic_stem_cell.htm (accessed January 30, 2008).

Kiessling, Ann A., and Scott Anderson. 2003. *Human Embryonic Stem Cells: An Introduction to the Science and Therapeutic Potential.* Sudbury, MA: Jones and Bartlett Publishers.

Lampman, Jane. 2001. "Different Faiths, Different Views on Stem Cells." *Christian Science Monitor,* July 23. http://www.csmonitor.com/2001/0723/p1s2.html (accessed January 30, 2008).

Larsen, W. J. 2001. *Human Embryology.* 3rd ed. New York: Churchill Livingstone.

Monroe, K. R., R. B. Miller, and J. Tobis, eds. 2008. *Fundamentals of the Stem Cell Debate: The Scientific, Religious, Ethical, and Political Issues.* Berkeley: University of California Press.

Morowitz, H. J., and J. S. Trefil. 1992. *The Facts of Life: Science and the Abortion Controversy.* New York: Oxford University Press. See, especially, 105–25.

Panno, Joseph. 2005. *Stem Cell Research: Medical Applications and Ethical Controversy.* New York: Facts On File.

Peters, T., and G. Bennett. 2003. "A Plea for Beneficence: Reframing the Embryo Debate." In *God and the Embryo,* edited by B. Waters and R. Cole-Turner. Washington, DC: Georgetown University Press.

Pontifical Academy for Life. 2000. "Declaration on the Production and the Scientific and Therapeutic Use of Human and Embryonic Stem Cells." Vatican City. http://www.vatican.va/roman_curia/pontifical_academies/acdlife/documents/rc_pa_acdlife_doc_20000824_cellule-staminali_en.html (accessed January 30, 2008).

Presbyterian Church (USA). 2001. "Attachment A: Statement on the Ethical and Moral Implications of Stem Cell and Fetal Tissue Research." Actions of the 213th General Assembly from the Office of the General Assembly. http://www.pcusa.org/oga/actions-of-213.htm#attachment (accessed January 30, 2008).

President's Council on Bioethics. 2005. "White Paper: Alternative Sources of Human Pluripotent Stem Cells." Washington, DC. www.bioethics.gov/reports/whitepaper/fulldoc.html (accessed January 30, 2008).

Sachedina, Abdulaziz. 2000. "Islamic Perspectives on Research with Human Embryonic Stem Cells." In *Ethical Issues in Human Stem Cell Research.* Vol. 3, *Religious Perspectives.* Rockville, MD: National Bioethics Commission.

Sandel, M. J., ed. 2007. *Justice: A Reader.* New York: Oxford University Press.

Science Daily. 2007. "Human Stem Cell Transplants Repair Rat Spinal Cords." February. http://www.sciencedaily.com/releases/2007/02/070213142747.htm (accessed January 30, 2008).

Snow, Nancy E., ed. 2003. *Stem Cell Research: New Frontiers in Science and Ethics.* Notre Dame, IN: University of Notre Dame Press.

Southern Baptist Convention (SBC). 1999. "Resolution #17: Human Embryonic and Stem Cell Research." Adopted at the SBC on June 16. www.johnstonsarchive.net/baptist/sbcares.html (accessed January 30, 2008).

Thomson, J. A., J. Liskovitz-Eldor, S. S. Shapiro, M. A. Waknitz, J. J. Swiegiel, V. S. Marshall, and J. J. Jones. 1998. "Embryonic Stem Cell Lines Derived from Human Blastocysts." *Science* 282: 1145–47.

Tooley, Michael. 1983. *Abortion and Infanticide.* New York: Oxford University Press. See, especially, 51.

Vogel, Gretchen. 1999. "Breakthrough of the Year: Capturing the Promise of Youth." *Science* 286 (5448): 2238.

Walters, L. 2004. "Human Embryonic Stem Cell Research: An Intercultural Perspective." *Kennedy Institute of Ethics Journal* 14: 3–38.

Waters, B., and R. Cole-Turner, eds. 2003. *God and the Embryo: Religious Voices on Stem Cells and Cloning.* Washington, DC: Georgetown University Press.

Weiss, R. 2005. "The Power to Divide." *National Geographic* 208 (1): 2–27.

Wilmut, I., and J. Taylor. 2007. "News & Views: Primates Join the Club." *Nature* 450: 485–86. This nontechnical summary of the technical report on therapeutic cloning in monkeys describes the significance of the accomplishment for eventual treatment of human disease and injury by therapeutic cloning.

3
Embryo Selection
Preimplantation Genetic Diagnosis

She's a typical teenage girl, she loves to dance. . . . We never thought
she would live to see 15. . . . Adam knows he helped his sister, that's all.
They're normal kids.

—Lisa Nash, interview by Dan Vergano for *USA Today*

Prospective parenthood can be anxiety ridden when one or both partners
carry a genetic disorder in their DNA or advancing age puts children at risk
for genetic abnormalities. If possible, most persons would choose not to
bring a child into the world who has little or no chance of living to adulthood or whose life will be filled with suffering. Genetic counselors examine parents' ancestries and assess the risk for concerned parents. With
enough information, some parents opt to use donor eggs or sperm to avoid
inherited abnormalities. This chapter is not about that option. Rather, it is
about a rather new option for genetically at-risk parents who embark upon
parenthood with their own eggs and sperm, *preimplantation genetic diagnosis (PGD)*.

PGD is technology that allows parents to assess the genetic health of embryos generated by *in vitro fertilization (IVF)* (see fig. 2.3). A major advantage
of PGD over traditional diagnostic procedures like amniocentesis is that
genetic diagnosis occurs before embryos are ever in the womb. By selecting only genetically healthy embryos for implantation, prospective parents
greatly increase the odds of having a healthy baby. But PGD carries with it
several ethical issues. This chapter addresses the following questions about
PGD and the challenges it raises for human values:

1. What are traditional prenatal screening/diagnostic tests for at-risk parents and how does PGD differ from these?
2. How is PGD currently used?
3. Will PGD usher in a new era of human *eugenics*?
4. Will PGD sometimes bring personal values of parents and physicians into conflict?
5. What is a *savior sibling,* and is it ethical to use PGD to create savior siblings?
6. Do IVF and PGD pose risks to women, fetuses, or children?
7. How might PGD affect families and society?
8. Should the use of PGD be regulated?

Preimplantation Genetic Diagnosis: The Biology

An array of prenatal tests is available to assess the genetic health of developing embryos and fetuses *in utero.* Noninvasive ultrasound and blood screening tests performed between ten and eighteen weeks of pregnancy assess the baby's risk of having congenital heart defects and chromosomal abnormalities, including the one that causes Down syndrome. But these tests are not diagnostic. For a diagnosis, couples need to use invasive tests: chorionic villus sampling (CVS) near the end of the first trimester or amniocentesis midway through the second trimester.[1] Results from these tests help couples to prepare for a child with special needs and to weigh the good and ill of terminating a pregnancy.

When conception is by IVF, PGD offers another option. Since the early 1990s PGD has made it possible for couples to obtain genetic information about their embryos while they are still small clumps of cells growing in a laboratory dish. PGD and embryo selection raise ethical issues for parents, doctors, and society that are different from those raised by prenatal screening and intrauterine diagnostic tests.

Preimplantation Genetic Diagnosis: What Is It? Why or Why Not Do It?

Available since 1990, PGD allows genetic diagnosis of IVF-conceived embryos when they are a small ball of six to ten cells growing in the laboratory. The procedure allows parents to cull defective embryos and to se-

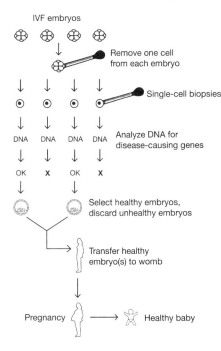

Fig. 3.1. Preimplantation genetic diagnosis by single-cell biopsy. Embryos produced by IVF are tested for genetic disease by analyzing the DNA extracted from a single cell removed from each embryo at the six- to ten-cell stage. Removal of one or two cells at this stage does not adversely affect the development of the embryo. Disease-free embryos are transferred to the uterus or stored frozen for later transfer. Embryos with genetic defects are discarded or donated by the parents for research.

lect only embryos with or without certain genetic traits for transfer to the uterus.

The PGD technologist starts with a dozen or so embryos obtained by IVF from a couple at risk for passing genetic abnormalities to their children. Two to three days after fertilization, a single cell is removed from each embryo without injuring the remaining portion of the embryo. DNA is extracted from each single-cell biopsy and subjected to specific genetic tests from among the hundreds now available for known genetic diseases and conditions (figs. 3.1 and 3.2).

Meanwhile, each biopsied embryo continues to grow and retain the potential to develop into a fetus if transferred to the woman's uterus. Depending on the results of DNA testing, certain embryos are selected for uterine transfer or stored frozen for future pregnancies. As for the genetically abnormal embryos, parents decide whether to discard them or donate them for research.

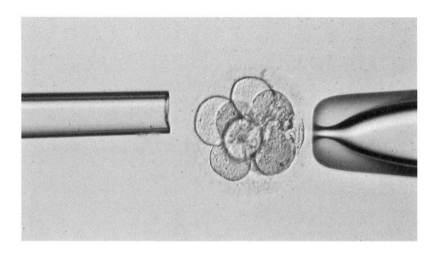

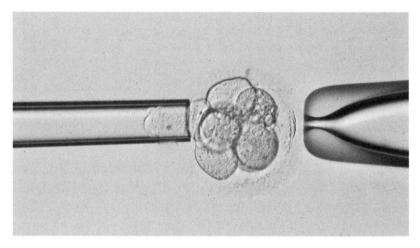

Fig. 3.2. Single-cell biopsy from a three-day-old human embryo during preimplantation genetic diagnosis. A slender, hollow, glass pipet approaches the embryo from its left side (top). A single cell removed from the embryo by suction through the pipet (bottom) will be biopsied. Its DNA will be amplified by polymerase chain reaction (PCR) technology and analyzed for specific genetic defects, while the remaining cells in the embryo continue to grow and can ultimately be transferred to the uterus to produce a fetus. The object to the right of the embryo in both photographs is a glass suction device for holding the embryo in place during biopsy. Photographs are by James J. Stachecki, Innovative Cryo Enterprises LLC/Natural World Photography, Linden, NJ, and are used with permission.

In 2005, a 97–99 percent success rate of obtaining cell biopsies for PGD was reported, and a diagnosis of genetically healthy or abnormal was obtained for 86 percent of the biopsied cells. Also in 2005, Harvard physician and bioethicist Sigal Klipstein (2005) noted that PGD biotechnology is ushering in a new era of predictive medicine that may ultimately supersede curative and preventive medicine in effectiveness and as a first choice in humankind's battle against disease. A major benefit of PGD is that couples at risk for having children with a genetic disease can bring to term a disease-free baby without resorting to prenatal, in utero diagnosis and pregnancy termination.

How Is PGD Now Used?

Currently, the most prevalent use for PGD is to prevent the birth of children with lethal or debilitating genetic diseases. In rare instances PGD is used for gender selection or to produce savior siblings, both discussed later in this chapter. Genetic diseases commonly diagnosed by PGD include cystic fibrosis, Tay-Sachs disease, Duchenne's muscular dystrophy, and Down syndrome. PGD is also used to diagnose late-onset diseases like Huntington disease and to identify embryos that would produce individuals with genetic predispositions for Alzheimer's disease and certain forms of cancer. Table 3.1 lists some of the many diseases and genetic conditions detectable by PGD.

The great power of PGD to prevent births of maldeveloped infants will soon be matched by equally extraordinary predictive powers for both disease- and non-disease-causing genetic conditions. DNA *base sequences* (i.e., the order of A, T, G, and Cs) are known for over twelve thousand genes that are inherited at predictable frequencies.[2] Several hundred of these genes are linked to specific disease and non-disease traits (morphological, behavioral, and physiological), and more such linkages are discovered weekly. Theoretically, PGD could be used to select for the presence or absence of any one of over twelve thousand genes, giving the procedure the potential for an almost unfathomable expansion into the realm of

Table 3.1. Some genetic diseases and conditions diagnosed by PGD

Disease	Inheritance	Symptoms; treatment	Diagnostic tests
Cystic fibrosis	Monogenic autosomal,[a] recessive[b]	Serious respiratory and digestive problems; antibiotics, double lung transplantation	PCR (a DNA test)
Duchenne (pseudo-hypertrophic) muscular dystrophy	Monogenic autosomal, recessive	Muscle weakness and wasting, including heart and breathing muscles; no cure	PCR
Sickle-cell anemia	Monogenic autosomal, recessive	Sickle-shaped red blood cells, ischemia, pain, fever, organ damage; bone marrow transplantation in children	PCR
β-thalassaemia	Monogenic autosomal, recessive	Anemia; blood transfusions, bone marrow transplantation	PCR
Tay-Sachs disease	Monogenic autosomal, recessive	Deafness, blindness, swallowing difficulty, muscle atrophy, paralysis; no cure or treatment	PCR, enzyme assay
Spinal muscular atrophy	Monogenic autosomal, recessive	Muscle weakness, atrophy; no cure, assistive technology includes ventilators and wheelchairs	PCR, electro-myography, muscle biopsy
MCAD	Monogenic autosomal, recessive	Inability to metabolize fats, seizures, apnea, cardiac arrest; low-fat diet	PCR, blood and urine tests

Disease	Genetic basis	Symptoms and treatment	Detection method
Congenital adrenal hyperplasia	Monogenic autosomal, recessive	Excessive or deficient sex steroids, ambiguous genitalia in females, precocious or delayed puberty; hormone replacement, genital reconstructive surgery	PCR, blood and urine tests
Huntington disease	Dynamic mutation,[c] autosomal, dominant[d]	Lack of coordination, declining mental abilities, dementia, usual onset 35–44 years; no cure, palliative treatments only	PCR
Turner syndrome	Monosomy X, all or portion of one X-chromosome missing in females	Short stature, webbed neck, sterility, cognitive and cardiovascular abnormalities; no cure, hormone replacement relieves some symptoms	Karyotype[e] analysis after PGD, CVS, or amniocentesis
Down syndrome	Trisomy in chromosome 21	Cognitive and growth deficits, almond-shaped eyes, congenital heart disease; no cure, management may include corrective surgery	Karyotype analysis after PGD, CVS, or amniocentesis
Lesch-Nyhan syndrome	X-linked,[f] recessive	Uric acid accumulation, gout, poor muscle control, moderate mental retardation; no cure, drugs can relieve some symptoms	PCR
Myotonic dystrophy type 1 (Steinert's disease)	Dynamic mutation, autosomal, dominant	Muscle weakness in extremities, cognitive problems; no cure, few treatments available for symptoms	PCR

Continued on the next page

Table 3.1. Continued

Disease	Inheritance	Symptoms; treatment	Diagnostic tests
Fragile X syndrome	Dynamic mutation, autosomal, dominant	Cognitive deficits, elongated face, large ears, poor muscle tone, social anxiety; no cure, behavioral therapy relieves some symptoms	PCR

[a] An autosome is a non-sex chromosome. Humans have twenty-two autosomes and two sex chromosomes, X and Y.

[b] Recessive refers to a mutation that must be present in both pairs of matched chromosomes in order to manifest itself. Thus, persons with only a single copy of a mutant, disease-causing gene are carriers of the disease condition but do not actually have the disease.

[c] A dynamic mutation changes (worsens) from generation to generation. Examples are trinucleotide repeats, sequences of three bases in chromosomal DNA that are repeated contiguously over and over. Huntington disease is caused by the repetition within human chromosome 4 of the trinucleotide CAG. The number of repetitions increases within each generation of carriers, and at a threshold level of repeats the disease manifests itself.

[d] Dominant refers to a mutation that only needs to be present in one member of a pair of chromosomes in order to manifest itself. Thus, persons with only a single copy of the gene variant that causes Huntington disease will have the disease.

[e] Karyotype refers to the number and appearance of chromosomes in a mitotic cell from a particular species. For example, the normal human karyotype is forty-six chromosomes, twenty-two pairs of autosomes (non-sex chromosomes) and one pair of sex chromosomes (XX in females and XY in males). One member of each pair of chromosomes comes from each parent. A missing chromosome (monosomy) or an extra chromosome (trisomy) causes syndromes like Turner (XO), Klinefelter (XXY), and Down (trisomy 21), which are diagnosed by a karyotype analysis.

[f] X-linked specifies a gene located on the X chromosome; most X-linked, recessive genetic diseases affect only males since they have only a single X chromosome. For females to be affected, both of their X chromosomes must carry the mutated, disease-causing gene.

embryo selection for reasons other than preventing the birth of babies with genetic abnormalities. This potential, coupled with our nearly certain future ability to genetically engineer human germ cells and early embryos (chapter 6), raises difficult ethical issues.

Preimplantaion Genetic Diagnosis: Human Values

One consequence of embryo selection via PGD is the rejection of some embryos for transfer into the uterus. No state or federal law in the United States prohibits discarding surplus or genetically abnormal embryos produced by IVF. Prospective parents using IVF and PGD bear sole responsibility for deciding what moral status to assign to their own embryos. Since the issue of personhood for embryos was thoroughly discussed in chapter 2, here we consider other ethical issues raised by PGD.

A triad of biotechnologies—IVF, *genetic engineering,* and PGD—ushers in an era of creativity for *Homo sapiens* of a type not previously experienced by our species in its two-hundred-thousand-year-old history. We are at the threshold of being able to transform ourselves according to our own concepts of biological perfection. PGD is the linchpin between current and imminent applications of human genetic engineering that will allow us to direct our own evolution.

The power to genetically reform ourselves via IVF, genetic engineering, and PGD did not appear overnight. IVF has been with us since the birth of Louise Brown, the first test-tube baby, on July 25, 1978. Recombinant DNA and DNA-sequencing technologies that make possible the genetic engineering of agricultural plants and animals also began in the mid-1970s. PGD itself was first used in the early 1990s. Completion of the Human Genome Project in 2001 and its spin-off discoveries and applications (chapter 4) are the glue that binds these three into a triad of transformative biotechnologies.

Having the human genome at one's fingertips in computer databases opens wide the door for developmental and molecular geneticists to identify a myriad of genes responsible for disease- and non-disease-causing traits that can be selected for or rejected via PGD. Therefore, eugenics is a major ethical issue associated with PGD. Some eugenic applications of PGD are already practiced, such as selection of the gender of children and selection of embryos to become savior siblings. Other moral issues associated with PGD include conflicting interests and duties of parents and physicians, certain biomedical risks borne solely by women and children, and the potential disruption of family dynamics and social structure. We turn first to eugenics.

PGD and Eugenics

Just what is eugenics, and are there instances where eugenic applications of PGD are desirable or at least allowable? The term *eugenic* was coined in 1883 by English scientist and writer Sir Francis Galton to refer to "all influences that tend in however remote a degree to give to the more suitable races or strains of blood a better chance of prevailing speedily over the less suitable than they otherwise would have had" (1883, 17 n 1). Actively selecting certain traits for future generations is called positive eugenics, while culling certain traits from the population is negative eugenics. Both types of eugenics have terrible legacies of racial and cultural discrimination and human rights abuse. Better Babies Contests and Fitter Family Exhibits at state fairs in the United States in the 1920s and 1930s and the National Socialist program in Nazi Germany that incentivized propagation of Aryan children are examples of positive eugenics. Forced sterilization of mentally ill or "deficient" persons in the United States and Canada in the 1920s and extermination of Jews and Gypsies during the Holocaust in Nazi Germany are extreme examples of negative eugenics.

Different people have different views about what makes for fine offspring, and these differences can harbor trouble. Judgments based on arrogance, ignorance, fear, hatred, and racism give eugenics its sordid history. The question now is whether eugenic measures using modern biotechnologies can or will be used to better human lives without discriminating against, oppressing, or otherwise harming others. Some political philosophers believe this is possible and advocate a so-called liberal eugenics con-

sisting of government-neutral, noncoercive genetic enhancements that respect the freedom and autonomy of both parents and children. Others rail
against departing from nature's traditional genetic lottery when it comes
to conceiving children.[3]

Human *functional genomics* correlates specific traits with specific genes.
Knowledge gained from functional genomics may someday allow individual parents to practice both negative and positive eugenics via PGD. By
choosing which embryos to discard and which to transfer to the uterus,
based on the embryos' genetic attributes and personal notions of what is
desirable and undesirable in offspring, parents could exert eugenic power
over the next generation. Human geneticists are still many years away from
a thorough understanding of the genetic bases for traits such as shyness,
gregariousness, cooperation, self-esteem, long- and short-term memory,
and cognitive skills used in mathematics, music, logic, literature, engineering, business, and art. But prudence now requires us to imagine a
day when genes for these and similar traits can be identified in embryos
by PGD and to begin preparing for the eugenic decisions that lie ahead.
Equally important will be legislative decisions affecting the availability of
PGD and identifying conditions whose prevention can be paid for using
public funds. For example, dwarfism and deafness are considered disabilities by some but not by others. Who will decide which genetic conditions
qualify as disabilities?

To a small degree, PGD already facilitates negative and positive eugenic
choices for a few parents who can afford the procedure. For example, embryos are sometimes discarded (negative eugenics) to avoid the birth of
babies with debilitating genetic diseases such as those in table 3.1.On the
other hand, embryos are sometimes selected (positive eugenics) to provide
stem cells that are immunologically compatible with a sibling in need of
stem cell therapy. We will return to the subject of eugenics and applications of modern biotechnologies in chapter 5 when considering human genetic enhancement. But now let's consider the specific case of using PGD
to select the gender of children.

PGD and Gender Selection

PGD easily identifies the gender of tested embryos.[4] Gender selection by PGD is currently used to prevent the birth of sons with X-linked diseases like hemophilia and muscular dystrophy (negative eugenics), and occasionally it is used for family balancing.[5] With the latter, both positive and negative eugenics apply since embryos of one gender are discarded while those of the other gender are selected.

For persons not opposed to IVF, gender selection by PGD for disease prevention is not controversial. But applying PGD in the absence of disease risk is another story. The prevalent position among bioethicists and within the medical community is that it is unethical to use PGD to select embryos of one gender over the other when disease is not an issue. What could be wrong with selecting the gender of your child?

The Council on Ethical and Judicial Affairs of the American Medical Association published a report in 1994 on ethical issues related to prenatal genetic testing. The authors view gender selection as dangerously discriminatory and warn that it could (1) foster the implicit valuing of one gender above the other, (2) lead to treating gender as a genetic disease, and (3) disrupt normal social relationships and encourage discrimination and violence against women (AMA 1994).

Gender should be considered a benign trait. In any society claiming to be egalitarian, one gender ought not to be valued above the other. But all societies do not claim to be gender blind. In countries including India and China, there is evidence that infanticide and gender-specific abortion are used to favor the birth of sons. If PGD were added to the repertoire of gender-selection methods in these countries, it could reduce the incidence of infanticide, on the one hand, but worsen the societal inequalities for females, on the other.

In countries where, under the law, society is egalitarian, gender selection is occasionally used for family balancing. Is this an ethically acceptable practice? Arguments against gender selection, even for family balancing, are based mainly on fears that the practice will ultimately lead to an implicit valuing of one gender above the other, to treating gender as if it were a genetic disease, and to favoring other non-disease-related traits. Ethicists warn that when nonmedical traits such as hair and eye color and musical or athletic ability become identifiable by genetic tests, parents may want to select for these.

Two arguments favor allowing gender selection by PGD. The first maintains that gender selection is just another parental reproductive right alongside the rights to decide whether, when, and how to have children. Parenthood already includes the right to nurture traits such as musicianship, athleticism, and scholarship with private lessons or tutors. Religiosity, humility, competitiveness, and other personality traits can be enhanced by example and exposure to certain environments. Selecting the gender of one's child, it is argued, is not qualitatively different from existing, sanctioned parental activities like these. The second argument for allowing gender selection is that family planning and balancing are legitimate uses for PGD that could actually reduce the human population and thereby relieve stress on the biosphere and on global society. For example, family sizes would decrease for couples wishing to have just one girl and one boy and who might otherwise have larger families due to a string of like-gendered children before finally having a baby of the other gender. Considering the vastly larger environmental footprint created by persons living in highly industrialized countries like the United States or England, where PGD is more apt to be used, reducing consumer numbers through reliable family-balancing methods could become a nontrivial contribution to planetary health.

Some ethicists worry that widespread gender selection will lead to devaluation of one gender, even in societies that strive for the equal treatment and valuation of all persons. Countering this is the notion that if parents came to prefer one gender slightly over the other, the rarer gender would increase in value, and preference would swing in the other direction. A low amplitude oscillation of parental preference between genders over time does not seem dangerous to society or to either gender. For those who fear that gender selection would favor male children, a 1970s poll of married women in the United States should be comforting. A majority of women polled preferred their firstborn to be a son and their second child to be a daughter, if the first was indeed a son. And women who already had a son and a daughter were evenly divided on gender preference for their third child (Westoff and Rindfuss 1974, 635).

No legislation prevents parents from using PGD for gender selection in the United States. However, responsibilities that physicians feel toward their profession usually discourage the use of PGD to select for benign traits like gender. The average cost of IVF with PGD in 2010 is over $15,000 and not usually covered by health insurance. So in the future, if using PGD for gender selection or to select embryos with certain other non-disease traits comes into vogue, only relatively wealthy couples will be able to afford it. Discrimination based on socioeconomic class could therefore become a major ethical issue associated with PGD.

Parents and Physicians: Autonomy and Conflicting Responsibilities

In the United States, assisted reproduction (AR) technologies, including IVF and PGD, are not monitored by the federal government. In Western society, parents have the freedom and autonomy to make reproductive decisions. Yet, a physician also has personal autonomy and professional responsibilities, a foremost one being to look after the welfare of the patient. In the case of PGD, who is the patient? Is it the parent(s) or the unborn child? Let's see how parental autonomy and physician responsibilities might sometimes conflict when a couple requests PGD.

A non-formalized human right in the United States and in most Western countries is the right for adults to reproduce. This is evidenced by the accessibility and acceptance of assisted reproduction technologies like IVF, gamete intrafallopian transfer (GIFT),[6] and intracytoplasmic sperm injection (ICSI).[7] With procreative autonomy come legal responsibilities to provide basic levels of nutrition, shelter, and education for children. Parents also have the freedom, within legal bounds, to do what they believe is best for their children. Beneficent parental care ranges from providing piano lessons and coaching soccer to arranging for growth hormone treatments and cosmetic surgery. As we saw with sex selection, and will examine later with savior siblings, parental autonomy can even include the use of AR technologies to further the interests of unborn children, preexisting children, and the family.

Physicians also have autonomy and responsibilities. The latter are encapsulated in documents like the Hippocratic Oath (2012), which describes the obligation to benefit the sick and do no harm. The American Medical Association (AMA 2001) and the World Medical Association (WMA 2007) have also adopted concise statements describing the proper physician-patient relationship. The AMA's "Principles of Medical Ethics" states that a physician's responsibility to the patient is paramount and that the physician "shall be dedicated to providing competent medical care, with compassion and respect for human dignity and rights" (2001, Principle I). Similarly, the WMA's "International Code of Medical Ethics" states that the duties of a physician include treating patients compassionately and respecting their human dignity, rights, and preferences. These documents also speak of the physician's responsibility to exercise his or her independent professional judgment (WMA 2006) and to contribute "to the improvement of the community and the betterment of public health" (AMA 2001, Principle VII).

Sometimes the autonomy, responsibilities, and personal values of parents and physicians come into conflict. A fascinating example of tension between parental autonomy and a physician's personal values arises when PGD is used to genetically test embryos for Huntington disease (HD) or other conditions that normally develop after childbearing years. Carriers of the HD trait have a 50 percent chance of passing it on to each of their children, but due to the late onset of the disease, carriers usually do not know they are carriers when they are ready to have children.[8] However, if an older sibling of a prospective parent has HD, the parent knows that he or she may be a carrier. An at-risk parent may have two strong desires: first, to have a future child free of the disease and, second, to remain uninformed about her or his own genetic condition for the disease. Both desires can be satisfied if the couple undergoes IVF and obtains a nondisclosing diagnosis of the embryos by PGD. With this arrangement, the clinic transfers only disease-free embryos to the mother's uterus and reveals nothing to the parents about the IVF or PGD results, including the number of eggs retrieved for IVF or the number of embryos examined by PGD.

An ethical dilemma for the physician arises if the first round of IVF or PGD fails to produce a pregnancy. This could happen if some IVF embryos carry HD and some are normal but development of the normal embryos becomes arrested before their transfer to the uterus. It could also happen if all embryos were normal, but none of them successfully implanted in the uterus. In the former case, the couple would simply be told that no embryos were available for transfer. They would not know whether this was because no eggs were retrieved, because all embryos were HD carriers, or because no embryos developed successfully *in vitro*. A nondisclosing diagnosis thereby preserves the desired ignorance of the at-risk parent about her or his own genetic condition.

On the other hand, if a large number (fifteen to twenty) of IVF embryos are obtained, all of which develop normally *in vitro* and lack the HD-causing gene, the physician can be sure that the at-risk parent is not an HD carrier. If transfer of a few of these embryos into the uterus fails to produce a pregnancy and the couple returns to the clinic for yet another round of embryo transplantation and PGD, the physician or clinicians are faced with an ethical dilemma. Should they tell the parents the good news that neither carries HD and that they have no reason to undergo IVF/PGD again? Doing this would relieve the woman from the risks and discomfort of IVF, not to mention the expense of yet another round of egg retrieval and/or embryo transfer. But giving parents this good news would violate the nondisclosing diagnosis agreement. What could be wrong with relaying good news? If parents were to know that they will be informed if they are free of HD, but then are told nothing, they will suspect that the at-risk parent is indeed a carrier, even though this is not an inescapable conclusion.[9] Telling parents good news even when they don't expect to be given such news is also problematic because patients using this rather rare application of PGD are in communication with similar patients via the Internet. Soon those not receiving good news would learn about the non-contractual practice of revealing good news, and by default they would deduce their likely HD status.

Ethical dilemmas can also arise when parents ask a physician for help in conceiving a child with one or more particular non-disease traits. If a physician's personal values tell him or her that the parents' request is not in the best interest of the unborn child, he or she is faced with having to choose between acting on his or her own values and respecting the procreative autonomy of parents. A rare example of this is requesting PGD to conceive a savior sibling to provide immunologically compatible stem cells or tissue for an existing family member needing a transplant. Who is the physician's patient in this case? Whose interest should be held paramount—the mother who must risk invasive procedures, the father, the unborn child, or the sick family member?

Savior Siblings

A famous savior sibling case is that of the Nash family from Denver in which parents used IVF and PGD to obtain a child whose umbilical cord blood was used as a source of *hemopoietic stem cells (HSCs)* to save the life of their daughter with a rare form of leukemia. Bioethicists have studied and written much about this case because it raises moral issues encountered with PGD that are now rare but could become more common in the near future.

In 1999, Molly Nash was five years old and suffered from Fanconi anemia, a rare genetic disease that causes leukemia. Molly was born with the disease. Both of her parents are genetic carriers of the disease, giving them a 25 percent chance of having a child with the disease, but they were unaware of this prior to Molly's birth. Molly could survive only if she received an HSC transplant from the bone marrow, peripheral blood, or umbilical cord blood of a compatible donor. Ensuring immunological compatibility between donor cells and the recipient is important in order to avoid rejection of donor tissues or organs by the recipient's immune system.

For children in Molly's position, the best source of HSCs is a genetically matched sibling, but Molly had no siblings. The next best source is from blood of a non-sibling, blood relative, but none of Molly's relatives were closely enough matched genetically to serve as a donor. The third best source is from imperfectly but sufficiently well-matched, unrelated donors registered with the National Marrow Program. For cases reported to the International Bone Marrow Transplant Registry between 1978 and 1994, Fanconi anemia patients' two-year survival rates after bone marrow transplants from identically matched sibling donors and from matched unrelated donors were 66 percent and 29 percent, respectively (Carreau 2006, 93). Faced with these statistics and Molly's certain death if a suitable donor was not found, the Nashes set out to use IVF and PGD to have a child, a savior sibling, whose umbilical cord blood could save their daughter (fig. 3.3).

The family used an IVF clinic in Denver and arranged for embryos at the six- to eight-cell stage to be flown to the Reproductive Genetics Institute in Chicago for PGD. Two tests were performed on one cell from each embryo, one for Fanconi anemia and the other for HLA (human leukocyte antigen) type.[10] The latter test identified embryos that would produce a child with HSCs compatible with Molly. In late 1999, after five rounds of IVF and PGD, at a cost of $100,000, PGD clinicians found one disease-free embryo with the proper HLA type. This embryo was transferred to the uterus of Lisa Nash, who gave birth to a son, Adam Nash, in August 2000. Cells from Adam's umbilical cord were frozen and flown to a University of Minnesota clinic in Minneapolis to await transplantation into Molly. After getting chemo- and radiation therapy to kill the leukemic cells in her bone marrow, Molly received Adam's cord blood cells, and by January 2001, Molly was cured of Fanconi anemia.

What new ethical issues arose when the Nashes used the "stacked" technologies of IVF, PGD, and stem cell transplantation (fig. 3.3) to create a savior sibling? Bioethicist Jeffrey Kahn and law professor Ana Mastroianni (2004) identify at least five such issues: (1) parents' motivation, (2) future physical and psychological risks to the donor, (3) a "slippery slope" toward embryo selection for other non-disease traits, (4) therapeutic uses of PGD, and (5) unintended consequences of the stacked technologies.

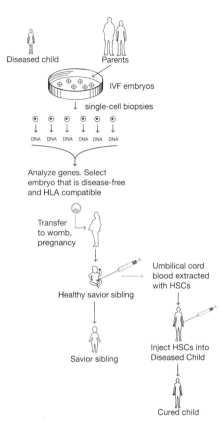

Diseased child

Parents

IVF embryos

↓ single-cell biopsies

DNA DNA DNA DNA DNA DNA

Analyze genes. Select embryo that is disease-free and HLA compatible

Transfer to womb, pregnancy

Healthy savior sibling

Savior sibling

Umbilical cord blood extracted with HSCs

Inject HSCs into Diseased Child

Cured child

Fig. 3.3. Creation of a savior sibling to provide hemopoietic stem cells (HSCs) to cure Fanconi anemia or other inherited leukemias. Parents provide many embryos by in vitro fertilization (IVF). The embryos undergo preimplantation genetic diagnosis (PGD), in which DNA from single cells is screened for disease-causing genes and analyzed to determine human leukocyte antigen (HLA) type. One or more disease-free embryos with an HLA type compatible with the sick sibling are transferred to the mother's womb. HSCs in blood from the umbilical cord of the resulting baby are transplanted into the diseased child to cure the condition.

When parents decide to have another child to treat a disease in an existing child, it may rightly be asked whether the savior sibling is simply being used as a means for the benefit of its sibling and parents. If this is the sole motivation for having the child, Immanuel Kant's moral dictum, treat all persons as ends in themselves and never simply as a means to achieve the ends of others, is violated. How does one assess the innermost motivations of parents? And even if parents of a savior sibling are deemed self-serving, one must acknowledge that other parents regularly have children for selfish reasons—to carry on the family name, provide an heir for a business, share work, mend a contentious marriage, or provide company for siblings.

Is it fair to scrutinize the motives of parents using PGD and embryo selection to produce a savior sibling without applying comparable scrutiny to the motives of all parents?

In the case of the Nash family, the savior sibling, Adam, appears to be loved and cherished for himself and not simply for having saved his sister's life. Preparing to return home after having spent four months in Minnesota monitoring Molly's post-transplantation progress, Jack Nash spoke for his family: "We are thrilled to see Molly return to the playful, energetic, young girl we love and appreciate. The bond already created between Molly and Adam is unlike any other sibling relationship, and we are truly grateful that they will be able to grow up and share a lifetime of experiences together—a lifetime that otherwise would not have been shared."[11]

Another famous savior sibling case, but not one involving PGD, is that of the Ayala family in California. Abe and Mary Ayala had a daughter, Anissa, with leukemia. In 1991 Anissa was in desperate need of a bone marrow transplant. Unable to find a donor with compatible cells, Abe had a vasectomy reversal operation, and the couple had a second child, Marissa, with bone marrow cells compatible with her sister. As adults, the sisters are very close, and Marissa is loved for herself, not just for saving her sister (Hardesty 2008).[12]

Still, there is no guarantee that all savior siblings would be treated like Adam Nash and Marissa Ayala. For his or her entire life, a savior sibling remains a genetically matched and potential tissue and organ donor for the older sibling. This is not a condition freely chosen by the savior sibling. A sibling saved by stem cells from cord blood may develop other conditions later in life that require a bone marrow transplant or even a kidney or lung donation, procedures that carry with them pain and physical risk for the donor. A savior sibling lives with the knowledge that he or she alone is the best donor for his or her sibling, and there are not yet any long-term studies on the psychological stress that this knowledge might cause. What it might be like to be a savior sibling is explored in a 2009 film, *My Sister's Keeper*, based on Jodi Picoult's 2004 novel with the same title.

Since HLA typing in an embryo examines a non-disease trait and thus provides no direct benefit for the child produced from that embryo, will HLA typing lead down a slippery slope to eventual testing for other non-disease traits such as physique, stature, or personality type? If so, is there anything wrong with using PGD to select for or against certain non-disease traits? Arguments against using PGD for embryo gender selection also apply to PGD used for other non-disease traits. For example, such selection could lead to devaluation of persons with unpopular traits, violate the autonomy of future persons, and encourage treating children as commodities. Some counter these arguments by noting that parents are already free to give their children advantages in life through music lessons, private schooling, travel, or tutors. Adding prenatal genetic advantages, they argue, should be viewed as just another manifestation of parental love and care. But these advantages would be available mainly to financially advantaged families. Over time, embryo selection through PGD could add a genetic component to societal divisions now based mainly on educational, cultural, and economic factors.

IVF and PGD: Special Issues for Women, Fetuses, and Children

Although PGD allows the genetic diagnosis of embryos before they are in the womb, it has some disadvantages absent for in utero tests. PGD is more expensive than CVS or amniocentesis,[13] and it poses unique risks for both mother and fetus.

IVF-associated risks begin when a woman's ovaries are stimulated with drugs or hormones to release one to two dozen eggs rather than the normal monthly ovulation of a single egg. Hyperstimulation syndrome can result, producing hypertension, mood changes, nausea and vomiting, abdominal pain, and, in 1 to 2 percent of cases, kidney or ovary damage and blood-clotting disorders. Egg retrieval itself, usually done under anesthesia, carries the risks associated with any invasive medical procedure. After embryo transfer, there are risks of an ectopic pregnancy and, when two or more embryos are transferred, a potential multiple pregnancy.[14] The latter increases rates of preterm labor, miscarriage, premature birth, cerebral palsy, and preeclampsia, a life-threatening, hypertensive disorder affecting both the mother and the fetus. In addition to these physical risks associated with IVF, women bear most of the psychological burdens associated with PGD and other types of prenatal genetic testing. No matter how much decision sharing occurs, it is ultimately the woman's mind and body that will receive or reject embryos, keep or abort fetuses.

Given these risks, why would a couple opt for PGD? Presently, the major reasons are (1) PGD's great diagnostic power and (2) removal of the specter of *elective abortion*. Using PGD for purposes other than disease prevention is on the near horizon, giving parents a new dimension of procreative liberty. At the same time, PGD poses some new ethical issues for families and society.

Family and Societal Dynamics

Use of PGD to select for benign, non-disease traits is still rare, occurring only when creation of a savior sibling is attempted or when a clinic agrees to use the technology for gender selection in family balancing. Already mentioned are the possible negative effects that expectations for later bone marrow or organ donation could have on the psyche of a savior sibling. But what if PGD's future use includes selection for traits like stature, physique, musical or mathematical capacities, gregariousness, or certain cognitive abilities? Knowledge that parents went to great lengths to help ensure a genetic predisposition for traits like these could put undue pressures of expectation upon children. And what if one sibling was chosen through embryo selection, or genetically customized while yet an embryo, to possess certain traits, while another sibling was not? Would the child conceived in the natural way feel less valued or genetically inferior to the other? Parental disappointment over children not meeting their expectations is not new, but widespread genetic commodification of children would be a new phenomenon with unknown consequences for family dynamics.

Although no negative, society-wide consequences of PGD are currently evident, several writers warn that the future use of PGD to select for benign traits, or to obtain embryos genetically manipulated to possess such traits, could produce divisiveness and inequities in society. The Council on Ethical and Judicial Affairs of the American Medical Association (AMA 1994) and other writers warn that PGD could ultimately produce a two-tiered society. In this scenario, children of a relatively wealthy group of individuals are free of genetic disease and possess predispositions for desirable physical, cognitive, and personality traits; whereas, genetic disease and less desirable traits accumulate in the lower tier of the population whose members cannot afford PGD or genetic enhancement technologies. Even if such a tiered society did not develop, there is danger for marginalization and inattention to those with genetic diseases or disabilities in a society that is systematically using genetic and assisted reproduction technologies to eliminate undesired genes from the gene pool. Persons carrying these genes would likely be less attractive as mates and perhaps be looked upon as having been products of irresponsible parents who did nothing to prevent their birth.

Social pressure on parents to employ expensive genetic and reproductive technologies to give their children a head start could create financial burdens on many families and decrease appreciation for what ethicist Michael Sandel (2007) calls the "giftedness" of children conceived naturally. Germ-line genetic enhancements via PGD could make physical and psychological characteristics of individuals in one generation subject to parental values and the whims of social preferences of an earlier generation. Finally, our appreciation for diversity, achieved through the blood and suffering of many generations, could evaporate with increased pressure for social uniformity through genetic biotechnologies.

Regulation of PGD

Presently IVF and PGD in the United States are not subject to governmental regulation. According to an issue brief in the *Duke Law and Technology Review* (Roberts 2002), one reason for this is lobbying by anti-abortion groups in the United States that keeps in place the ban on federal funding for institutions doing research with human embryos and assisted conception. The result is that IVF and PGD are not performed in government-funded universities or hospitals. Instead, the market drives these procedures into the private sector and encourages business-oriented approaches to human embryo research and the development of assisted reproduction technologies.

Attempting to fill the void in oversight of assisted reproduction procedures and practices at private clinics in the United States, the Society for Assisted Reproductive Technology (SART) requires accreditation of its member laboratories. But membership in SART is voluntary, and many clinics are not members. The Ethics Committee within the American Society of Reproductive Medicine has produced guidelines for assisted reproduction practices, but membership in this organization is also voluntary. The American Medical Association suggests guidelines for the future genetic manipulation of human embryos that are also applicable to the use of PGD for the selection or rejection of specific non-disease-causing genetic traits, but these are nonbinding. As for state legislation regulating PGD, there is almost none. At this writing, laws in just four states restrict PGD by mandating that it must do no harm to the embryo and must result in some benefit to the resulting individual(s).

In contrast to the situation in the United States, the United Kingdom established the Human Fertilization and Embryology Authority (HFEA) in 1990 to license and regulate assisted reproduction clinics. To ensure broad public representation, half of the members of the HFEA must be persons not associated with medical professions. One responsibility of the HFEA is to define boundaries for treatments and research involving human embryos. Government's relationship with PGD in other countries varies widely from prohibiting it entirely (Switzerland, Austria, and Germany) to not addressing it at all as in the United States.[15]

Conclusion

IVF, coupled with PGD, allows parents the option of not having children with certain genetic diseases or of having children with certain non-disease traits without resorting to abortion. A diversity of ethical and social issues is associated with PGD. Some of these, such as the accessibility of the technology and the moral status of the embryo, are not unique to PGD. Applications of PGD that bring up special ethical issues include gender selection and creation of savior siblings. PGD can also create tension between parental procreative autonomy and physician responsibility. The daily expansion of human genomic information for disease and non-disease traits and advances in genetic manipulation techniques portend an ever-increasing eugenic potential for PGD. A national dialogue and regulation of PGD and other assisted reproduction procedures lag far behind these rapidly advancing technologies in the United States. The Human Fertilization and Embryology Authority in the United Kingdom provides a model for formalized public input into government oversight and regulation of procedures like IVF and PGD.

Questions for Thought and Discussion

1. Do you believe it is all right to use embryo selection by PGD to prevent the birth of babies with genetic diseases? If so, should PGD be restricted to disease prevention; or should couples be able to use PGD to select embryos with certain non-disease traits for implantation? Should there be any legal restrictions on the use of PGD? If so, what restrictions do you suggest?

2. Suppose that you and your spouse are ready to have children and have just learned that one of you carries a form of the gene for Huntington disease (HD) that does not affect the carrier but is likely to cause lethal disease in 50 percent of your children during their midlife.[16] Both of you have strong personal religious beliefs that personhood begins at conception; you have just learned how IVF and PGD offer the possibility of having a normal child. What will you do, and how do you arrive at your decision?

3. Suppose that you and your spouse are committed to having no more than three children and that you both strongly desire to have at least one boy and one girl. You already have two boys. Presuming cost is not an issue, will you use IVF and PGD to select the gender of your third child? If not, why not?

4. Suppose you are the head of an IVF or PGD clinic. If the couple in question 3 requests your services, will you accept them as clients?

5. You are a married parent with a two-year-old daughter with acute lymphoblastic leukemia; your daughter needs a stem cell transplant in order to live more than six years, and no compatible donors have been found within the family or in the national stem cell bank. You are considering IVF and PGD to have a savior sibling whose umbilical cord cells can be used to save the life of your daughter. For you personally, what are the pros and cons of taking this route to save your daughter?

6. If PGD becomes subject to legislative regulations, should the regulation be at the federal, state, or local level?

Notes

1. CVS involves extracting cell samples from small fingerlike projections of the chorionic villus, a component of the placenta formed by the embryo itself. Since the genetic makeup of chorionic villus cells matches that of the fetus, CVS and genetic analysis of the sampled cells reveal whether the fetus carries disease-causing genes. Except for spinal cord and brain defects, virtually all of the conditions testable by amniocentesis can be detected by CVS.

2. Mendelian inheritance of a gene occurs when particular forms (alleles) of the gene follow predictable patterns of segregation into eggs and sperm and thereby result in predictable ratios of offspring that carry traits specified by those alleles.

3. Ethicist Michael J. Sandel (2007), in his book *The Case Against Perfection,* expresses fear that the state might eventually compel parents to select for certain "general-purpose" enhancements in their offspring, such as intelligence, and thereby tilt eugenics back toward the bad, government-sponsored eugenics of the past. He also argues that genetically designing children threatens the social practice of parenting with unconditional love. Sandel also nicely summarizes the views of proponents of "liberal eugenics," including Nicholas Agar, Allen Buchanan, Dan W. Brock, Norman Daniels, Daniel Wikler, Ronald Dworkin, and Robert Nozick.

4. The gender of an embryo is determined by karyotyping cells derived from a single cell in the embryo. The biopsied cell is grown in laboratory culture; growth of its cell offspring are arrested in mitosis, when the chromosomes

are highly condensed and easily identified and counted using a microscope. Cells with two X chromosomes come from female embryos and those with a single X and single Y chromosome indicate maleness.

5. If a disease-causing gene is carried on the X chromosome, the disease is said to be "X-linked." A female human may have a disease-causing gene on one of her X chromosomes but not on the other. If the gene is recessive, the single normal X chromosome will prevent the disease from manifesting itself. In this case, the woman is a carrier of the disease but does not have the disease herself. However, since 50 percent of her eggs will contain a disease-causing X chromosome, 50 percent of her sons will have the disease. This is because males possess only a single X chromosome per cell.

6. GIFT is a variation on IVF in which eggs and sperm harvested from prospective parents are injected into the woman's oviduct, where fertilization occurs in its natural environment instead of in a glass dish in the laboratory.

7. In ICSI, fertilization is accomplished by direct injection of a sperm into the egg's cytoplasm. It is used when the male partner has such low sperm numbers in his ejaculate that fertilization after intercourse is highly unlikely.

8. Although the mutation causing Huntington disease (HD) is a dominant mutation, a person can be a carrier of the trait without actually manifesting the disease. This is because the mutation consists of a series of trinucleotide repeats in the HD gene. The number of repeats determines whether disease symptoms will occur, and usually the number of repeats increases in each successive generation of carriers.

9. Reason for failure of the first IVF attempt could be due to the unlikely event that no eggs were retrieved or that HD-free embryos died before uterine transfer.

10. HLA (human leukocyte antigen) is a protein indicator of genetic matching between the patient and potential donors of cells, tissues, or organs.

11. University of Minnesota News Release, January 4, 2001, "Umbilical Cord Transplant Succeeds for Molly Nash." http://www1.umn.edu/news/news-releases /2001/UR_RELEASE_MIG_1251.html (accessed June 7, 2012).

12. The baby born to save her sister says she has no regrets. *The Orange County Register,* July 21. In this interesting news story seventeen years after the birth of

Marissa, the sisters hint at the possibility that Marissa may someday donate an egg so that her sister can have a child by IVF. As it is, Anissa is sterile due to earlier radiation treatment for her leukemia. http://www.ocregister.com/articles /marissa-anissa-ayala-2100465-marrow-story# (accessed July 4, 2009).

13. IVF, which is necessary for PGD, costs $10,000 to $14,000 per attempt and usually is not covered by insurance.

14. In ectopic pregnancies, the embryo implants in a Fallopian tube (95 percent of the time) or in the ovary, cervix, or abdomen. An ectopic pregnancy never leads to a live birth, and if the embryo or fetus is not removed early enough, it can cause the organ containing it to rupture and threaten the life of the mother.

15. World Medical Association's (2006) "International Code of Medical Ethics" was adopted by the 3rd General Assembly of the World Medical Association, London, England, October 1949, and amended by the 22nd World Medical Assembly, Sydney, Australia, August 1968, and the 35th World Medical Assembly Venice, Italy, October 1983, and the WMA General Assembly, Pilanesberg, South Africa, October 2006. http://www.wma.net/e/policy/c8.htm (accessed March 28, 2007).

16. This form of the HD gene has what is called "reduced penetrance." The genetic abnormality is not great enough to cause HD in its carrier, but during sperm (or egg) production, the abnormality worsens so that it is likely to cause the disease in children.

Sources for Additional Information

American Medical Association (AMA). 1994. CEJA Report D-I-92, "Prenatal Genetic Screening." http://www.ama-assn.org/resources/doc/code-medical-ethics /212a.pdf (accessed June 7, 2012).

Garreau, J. 2005. *Radical Evolution: The Promise and Peril of Enhancing Our Minds, Our Bodies—and What It Means to Be Human.* New York: Doubleday.

Gilbert, S. F., A. L. Tyler, and E. J. Zackin. 2005. *Bioethics and the New Embryology: Springboards for Debate.* Sunderland, MA: Sinauer Associates.

Hippocratic Oath. 2012. http://www.pbs.org/wgbh/nova/body/hippocratic-oath -today.html (accessed May 7, 2012). This site has both the classical version and the modern version written by Louis Lasagna in 1964.

Kahn, J. P., and A. C. Mastroianni. 2004. "Creating a Stem Cell Donor: A Case Study in Reproductive Genetics." *Kennedy Institute of Ethics Journal* 14 (1): 81–96.

Klipstein, S. 2005. "Preimplantation Genetic Diagnosis: Technological Promise and Ethical Perils." *Fertility and Sterility* 83: 1347–1534.

Knoppers, Bartha M., Sylvie Bordet, and Rosario M. Isasi. 2006. "Preimplantation Genetic Diagnosis: An Overview of Socio-Ethical and Legal Considerations." *Annual Review of Genomics and Human Genetics* 7: 201–21.

Prenatal Tests: An Overview (First, Second, and Third-Trimester Tests). http://www.babycenter.com (accessed February 13, 2007). This site has accurate information about prenatal tests, including the integrated prenatal screening test, amniocentesis, and chorionic villus sampling.

Principles of Medical Ethics Adopted by the AMA's House of Delegates June 17, 2001. http://www.ama-assn.org/ama/pub/category/2512.html (accessed March 28, 2007).

Roberts, J. C. 2002. "Customizing Conception: A Survey of Preimplantation Genetic Diagnosis and the Resulting Social, Ethical, and Legal Dilemmas." *Duke Law and Technology Review.* http://www.law.duke.edu/journals/dltr/articles/2202dltr0012.html (accessed June 7, 2012).

Sandel, Michael J. 2007. *The Case against Perfection.* Cambridge, MA: Belknap Press of Harvard University Press.

World Medical Association. 2007. "World Medical Association International Code of Medical Ethics." http://www.wma.net/e/policy/c8.htm (accessed March 28, 2007). This was adopted by the 3rd General Assembly of the World Medical Association, London, England, October 1949, and amended by the 22nd World Medical Assembly, Sydney, Australia, August 1968, and the 35th World Medical Assembly Venice, Italy, October 1983, and the WMA General Assembly, Pilanesberg, South Africa, October 2006.

4
The Human Genome Project

Spin-Offs and Fallout

I firmly believe that the next great breakthrough in bioscience could come from a 15-year-old who downloads the human genome in Egypt.
——Thomas Friedman

The deciphered human *genome* is called by some the Holy Grail for medicine and cell biology. Just what is a genome, and why is it important to learn about the human genome? In chapter 1 we saw that the combined *DNA* from one copy of each type of chromosome in an organism constitutes that species' nuclear genome (see fig. 1.8).[1] Determining the *base* (A, T, C and G) *sequence* in an organism's DNA, identifying the location of *genes,* and analyzing the organization of the DNA in *chromosomes* is called *genomics.* Knowing the precise base sequence of an organism's DNA allows genome researchers to identify the sequences coding for *proteins* and to deduce from the *genetic code* (chapter 1) the *amino acid* sequence of every protein required for the development and life of that organism.

In 1988 the United States made a national commitment to determine the sequence of all 3.1 billion base pairs of the human genome. The venture was dubbed the Human Genome Project (HGP). Soon after the United States committed to a HGP, other nations were brought on board as collaborators in an international public consortium, the International Human Genome Sequencing Consortium (IHGSC), to sequence the entire human genome. In June 2000, researchers announced that the HGP was virtually completed. In this chapter we consider some questions about the HGP and about the genomics of two of humankind's close relatives:

1. Why was the HGP undertaken, and how were metaphors important for its funding?
2. How was completion of the HGP spurred on by competition?
3. What beneficial spin-offs emerge from the HGP?
4. Where is human genome research going next?
5. What are the ethical, legal, and social implications of the HGP?
6. What is the epigenome?
7. How can information from the Chimpanzee and Neanderthal Genome Projects benefit humankind?

Human Genome Project: History and the Science

There is no better source of information about all aspects of the HGP than the project's official website, activated for researchers and the public in 1994.[2] Since then, it has been updated regularly, sometimes daily, and it has links to all of the important documents and publications associated with the HGP since its inception. Let's indulge ourselves with a sampling of some of the major events, objectives, and findings of the HGP and some of the media-reported human interest aspects of the project.

Selling the HGP to the Public

During the early 1980s molecular biologists recognized that the technology existed to determine the entire base sequence for the human genome. But the feasibility of actually doing it was in doubt for several reasons. First, estimates were that it would take a commitment from hundreds of scientists over a period of fifteen to twenty years to complete the project. At an estimated cost of $1.00 per base pair, the project would cost over $3 billion. Also, the usefulness of the project was controversial and uncertain. Nevertheless, the idea of charting the human genome continued to excite and intrigue many scientists. By 1986 the pros and cons of embarking upon the project were being hotly debated in the scientific community. Scientists in favor of the project stressed the potential medical benefits of learning more about the genetic basis of inherited diseases. Researchers opposed to the project pointed to the limited resources available for biomedical research and warned of the detrimental effects of diverting funds from existing cancer, pediatric, and other medical research in order to fund a HGP.

In the end, HGP supporters carried the day and won a political battle for congressional support of the equivalent to a Manhattan Project for biomedicine. Alice Dreger (2000) of Michigan State University describes how HGP advocates did this in an essay titled "Metaphors of Morality in the Human Genome Project." Dreger argues that through clever use of terms, including *genetic frontier, gene mapping,* and *bad genes,* supporters of the HGP made it patriotic to support the project and unpatriotic to oppose it.

Americans view themselves as conquerors of frontiers—first there was the New World, then the Western frontier, the Moon, close inspections of Mars and the outer planets, and Hubble Telescope views of the early universe. By successfully portraying the HGP as the means to conquer a scientific frontier that would bring health and happiness to Americans and others throughout the world, it became difficult for legislators to argue against it. "Gene mapping" conjured up an image of biologist explorers mapping new territories to be claimed for the national interest. The phrase "bad genes" was also used to garner political support for the HGP. One HGP proponent borrowed the inscription on the Statue of Liberty, editorializing that the HGP was necessary to aid "the poor, the infirm, and the underprivileged" who through no fault of their own had inherited bad genes. What red-blooded American who believes in equal opportunity for all could oppose a project that promised to rescue innocent victims from bad genes? In 1988, the US Congress voted to fund an HGP through the National Institutes of Health. Nobel Laureate James Watson became the project's first director and promised to deliver the complete sequence of the human genome by 2005 at a cost of $3 billion. Work began in 1990.

Specific Objectives of the HGP

One of Dr. Watson's first actions as director was to make the HGP an international project. He recognized the importance of the project being a venture of humankind rather than being owned by just one nation. His efforts resulted in the establishment and cooperation of twenty genome-sequencing centers in Britain, Germany, France, Japan, China, and the United States. These collaborating centers comprised the IHGSC and became known as the Public Consortium. In addition to assembling an international team of researchers, the IHGSC committed to making its genomic sequence data available to the entire world as soon as it was obtained. At the end of each day's work, the sequences deciphered that day were posted on the consortium's public website. The IHGSC succeeded brilliantly in modeling international scientific cooperation.

Francis Collins replaced James Watson as director of the Public Consortium in 1993. While director, Collins and his colleagues at the NIH and Department of Energy (DOE) prepared two detailed five-year plans for the project, one in 1993 and the other in 1998. Both plans were published in *Science* magazine (Collins and Galas, 1993; Collins et al. 1998) and are on the HGP information website.[3] That site lists six project goals:

1. Identify all of the approximately 20,000–25,000 genes in human DNA.
2. Determine the sequences of the 3 billion chemical base pairs that make up human DNA.
3. Store this information in databases.
4. Improve tools for data analysis.
5. Transfer related technologies to the private sector.
6. Address the ethical, legal, and social issues (ELSI) that may arise from the project.

Two other high-priority objectives for the project were having accuracy in sequencing and providing technology to identify spots in the genome that differ between individuals. For accuracy, the project set a standard of less than one error in every ten thousand bases sequenced. Remarkably, when the final draft of the genome was completed, the error rate was only about one in every one hundred thousand bases (IHGSC 2004).[4] As for differences between the genomes of individuals, the project was especially interested in spots where the base sequence varies by only a single base.

These are called *single nucleotide polymorphisms (SNPs),* and cataloging them is very helpful for finding disease-associated genes and in studying human diversity and evolution (chapter 5).

The HGP met all six of its objectives in record time and under budget. By 2003 the HGP was completed, and results for the first two objectives were reported in the journal *Nature* in 2004 (IHGSC 2004). The remaining four objectives were addressed regularly throughout the thirteen-year duration of the project.

The genomic information, licensed technologies to private companies, and research grants to the private sector that emerged from the HGP spawned a multibillion-dollar, international biotechnology industry. Finally, much attention was given to the ethical, legal, and social issues associated with human genomics, the subject of the second half of this chapter. Now let's see who actually did the work and how they approached the problem of sequencing 3.1 billion bases of DNA.

Who Sequenced the Human Genome, How Did They
Do It, and Whose DNA Did They Use?

Two groups sequenced the human genome: the Public Consortium (IHGSC) and Celera Genomics, a private company headed by molecular biologist and entrepreneur J. Craig Venter. The IHGSC was coordinated by the DOE and NIH. In addition to the six countries in the IHGSC, a dozen other countries established sequencing centers to help in the project. More than 2,000 scientists contributed data to the Public Consortium. Celera Genomics' sequencing work was done by approximately 250 scientists. The majority of these were employed by Celera, but a few others were from some American universities and institutions in Australia, Israel, and Spain.

Two approaches to genome sequencing. The Public Consortium and the Celera Genomics group used different strategies to sequence the human genome. The IHGSC used a method called hierarchical sequencing to decode the human genome, whereas the Celera Genomics group used a whole-genome shotgun-sequencing method.

In the hierarchical method, genomic DNA is cut into many large, overlapping pieces of about 150 million bases pairs each. Modern laboratory methods in molecular biology and computer programs are then used to determine the relationship of these pieces to each other within the chromosomes of living cells. Then the large pieces are broken randomly into small, overlapping fragments, the base sequence of each of these fragments is determined, and a computer program reconstructs the sequence of bases that comprised the original, naturally occurring genome.

The whole-genome shotgun method used by Celera skips the steps involving the long pieces of DNA and simply blasts the whole genomic DNA sample into tiny, randomly generated, overlapping fragments. These fragments are then sequenced by an automated process and aligned into the entire genome by computer programs.

The hierarchical method is slower, more expensive, and requires more manual labor than the shotgun approach. The shotgun method lends itself better to automated processing but is more prone to error due to the occurrence of short sequences of bases that are repeated many times at many locations in the genome.

The Public Consortium and the Celera group both used DNA derived from anonymous donors for their projects. The IHGSC used DNA from eight males of unknown ethnicity for their starting material, with one donor's DNA comprising 75 percent of the entire sample. Celera used DNA from two males and three females (one African American, one Asian Chinese, one Hispanic Mexican, and two Caucasians), with one donor's DNA comprising 66 percent of the entire sample.

Competitors race toward the finish and then cooperate. On June 26, 2000, President Bill Clinton, Prime Minister Tony Blair, Francis Collins, and J. Craig Venter announced the completion of a rough draft for the sequence of the 3.1 billion bases in the human genome. Their announcement was the culmination of a fantastic story of international resolve and cooperation, scientific competition, personal rivalry, and human pride.

During the first eight years of the project, the Public Consortium progressed methodically, steadily, and slowly toward the 2005 goal. Its emphasis was on accuracy and openness. Each day the consortium posted the details of its progress on a public website.

Meanwhile, Venter and his group at Celera (Latin for "quick") Genomics developed their new shotgun-sequencing method. Recognizing the method's potential application to the human genome and the profits to be made from deciphering and patenting the structure of disease-causing genes, Venter and his associates announced their intention to sequence the human genome within three years, four years ahead of the Public Consortium's scheduled completion date.

At first, most scientists in the Public Consortium did not take Venter's boastful announcement seriously, believing that the shotgun method could not produce the accurate results so important to the HGP. Nevertheless, Francis Collins responded to Venter's challenge and moved the target date for the consortium's completion of a first draft of the human genome forward to the spring of 2000. In March 2000, Venter removed all doubts about the power of his shotgun method. In the middle of the race to finish the human genome, Venter and his collaborators took time out to sequence and publish the genome of the fruit fly, *Drosophila melanogaster,* a favorite experimental animal of geneticists for over a century.

For a short time, it looked as though the Public Consortium's objective of making the complete, decoded, human genome open information for the entire world was in jeopardy and that Venter would be seeking patents on many human genes that his group had sequenced first. Then, at the urging of President Clinton, the two rivals met May 7, 2000, over pizza and beer in the townhouse of a DOE scientist and agreed to pool their data and make a joint announcement in June that between the two groups, a rough draft of the human genome had been completed. The two groups published their results the same week in the world's two premier science journals, *Science* and *Nature* (IHGSC 2001; Venter et al. 2001).

How Many Genes Are in the Human Genome?

The exact number of genes needed to make a human being is not known, but the latest estimate is about 22,000, slightly more than the 19,000 genes needed to make a worm and fewer than the 25,000 for a fruit fly. How a human being can be constructed with fewer genes than a fly is a puzzle. One possibility is that multicelled animals have most of their genes in common and that it takes only a small number of gene differences to produce large anatomical, behavioral, and cognitive differences. Another idea is that regulating when genes are active and inactive, especially during early development, is more important to the final outcome than the number of genes. Humans may simply have different or more complex gene regulatory processes than their multicelled relatives. Finally, major anatomical differences between species of animals may reflect ways that gene products, particularly proteins, interact with each other. Maybe humans just have a larger and more complex network of interacting proteins than fruit flies.

Cost and Time to Sequence the Human Genome: Past, Present, and Future

The HGP was completed under budget for $2.7 billion two years ahead of schedule. Since work began on the HGP in 1990, steady advances in automating DNA sequencing led to a collection of current procedures collectively called high-throughput DNA sequencing. Now millions of DNA fragments can be sequenced simultaneously at relatively low cost. An early version of high-throughput technology produced the sequence of the entire diploid genome of J. Craig Venter in 2007 (Levy et al. 2007) at a cost of $100 million.[5] Also, in 2007, high-throughput sequencing instruments sequenced the diploid genome of Nobel Laureate and first director of the HGP, James Watson. Watson's genome was sequenced in just four months at a cost of $1.5 million, less than 0.06 percent the cost of the HGP. Next-generation genome-sequencing technologies will probably make it possible for virtually anybody to have her or his genome sequenced rapidly and relatively inexpensively, perhaps in less than a day for $100 (Morrow 2009).[6] In fact, two new DNA-sequencing machines capable of deciphering a human genome in twenty-four hours are scheduled for commercial marketing in 2012 (Scudellari 2012).

Developing the technology to do such a remarkable feat is one thing. But gaining the knowledge to usefully interpret complete genomic data from a particular human being is quite another. In fact, one person present at a counseling session for Dr. Watson to discuss the meaning of twenty mutations in his genome associated with an increased risk of disease was profoundly impressed at how little the counselors were able to advise him.[7]

Benefits from the HGP and What's Coming Next

Aside from the invaluable raw genetic data about our species provided by the HGP, its direct and indirect spin-off benefits are innumerable. Two major classes of benefits are those from international collaboration and cooperation and those from applications of the technology and data generated by the project. Potential applications of HGP research include personalized medicine, disease detection and health risk assessment, drug design based on the molecular basis of diseases, gene therapy, energy sources from biofuels, environmental remediation, pollution detection, reduction of atmospheric carbon dioxide, a better understanding of human evolution and ancient human migrations, DNA forensics, matching organ donors with recipients, and agricultural products. Gene therapy and some other spectacularly imaginative applications of genomics are discussed in chapters 5 and 9, respectively.

The stunning success of the IHGSC shows that the scientific community and its "way of knowing," i.e., the process of science itself, transcends ethnic and sociopolitical differences. Researchers at twenty sequencing centers in six countries showed the world what an international network of collaboration toward a common goal can accomplish. Relationships and the communication networks established during the thirteen-year duration of the HGP not only continue but are expanding. Thousands of researchers with diverse national and cultural backgrounds, unified by the findings and organizational structure of the IHGSC, now collaborate in the genomic analysis of many other species of organisms. Together, they seek knowledge about how the human genome and other genomes function toward the goal of using genomic information to advance human health and well-being.

Personal genomics and personalized medicine. Personal genomics, a type of commercialized genome technology, came on the scene in 2008. It caters to individuals who seek answers about the ancient ancestry, disease risk, or genetic disease-carrying status of their personal DNA.[8] Technologies like high-throughput DNA sequencing and rapid-screening methods for specific variations in *DNA sequences* make personal genomics possible. The industry is in its infancy, and it is difficult to discern strengths and weaknesses of the several service providers simply by reading from their websites. One thing that personal genomics companies have in common is catchy names: 23andMe (referring to our twenty-three chromosomes), deCODEme, Knome (pronounced "know me"), and Counsyl (suggesting an offering of advice for carriers of genetic disease). The cost for information about your DNA varies from $99,500 for a whole genome sequence to $399 for ancestry data and analysis for a sampling of genetic variants associated with disease risks. Lacking in most personal genomics services so far is a coupling of the genomic information with professional genetic counseling about what the information really means in terms of living one's life or having children.

A long-term goal of the fledgling personal genomics industry is *personalized medicine*. Tailoring treatment and preventative practices to an individual's unique genomic constitution is the aim of personalized medicine. Ideally, personalized medicine will eliminate adverse side effects from drugs and facilitate prescription of the appropriate types, combinations, and dosages of drugs for an individual by discerning the person's likely response to medicines before they are administered. Personalized medicine is coming, but currently there are not enough data from individual genomes representing a broad enough array of health conditions and responses to drugs to make it practical. Dr. George Church of Harvard Medical School in Boston is trying to remedy this situation by integrating genomic data, health status, family history, and environmental factors for one hundred thousand volunteers in the Personal Genome Project.[9]

As technologies continue to reduce the time and cost for sequencing individual human genomes, the feasibility of using a patient's whole-genome sequence to optimize treatments for disease, especially cancers, is increasing. In fact, some doctors in research hospitals are already using whole-genome sequencing to identify genetic variants in genes or in DNA regions that regulate the activity of genes in their patients. They use this information to help design the most effective treatments for disease and also to prevent disease and improve the overall health of their patients (Drmanac 2012).

Regenerative medicine. Another area of human medicine benefiting from HGP information is *regenerative medicine.* We saw in chapter 2 how stem cells may someday be used to regenerate diseased and injured tissues. Genomic information gleaned from the HGP will help identify genes coding for proteins that direct tissue repair and defend the body against viral and bacterial invaders. Pharmaceutical researchers can use this information to develop drugs that enhance the action of these proteins and to produce the proteins themselves for treating persons recovering from strokes, heart attacks, spinal cord injuries, and diseases like AIDS.

The Epigenome

Now that the human genome is deciphered, many genome researchers are turning their attention to the human *epigenome.*

Epi means "on the outside of." Thus, the epigenome is a layer of genetic information superimposed over the genome itself; that is, over the sequence of the A, T, G, and C bases in the DNA. Specifically, the epigenome consists of patterns of chemical modifications (*epigenetic* factors) to certain bases in the DNA or on some of the proteins associated with DNA molecules (fig. 4.1).[10]

One science journalist likens the epigenome to "chemical amendments that dangle like charms on a bracelet from the linear string of letters that spell out the genetic code."[11] To decode the meaning of those chemical amendments, a consortium of institutes and researchers recently organized the Human Epigenome Project (HEP).[12] Why is learning about the human epigenome so important?

A cell's epigenome regulates the activity of the genome. Since the genome's activity is different in different cell types, there must actually be many types of epigenomes. For example, there is an epigenome peculiar to liver cells, heart cells, nerve cells, gut cells, and to each type of cancer cell. The stated goal of the HEP is "to identify all the chemical changes and

The Epigenome

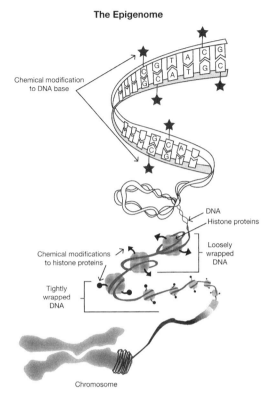

Chemical modification to DNA base

DNA
Histone proteins

Loosely wrapped DNA

Chemical modifications to histone proteins

Tightly wrapped DNA

Chromosome

Fig. 4.1. The epigenome consists of chemical modifications to some of the C bases that lie next to Gs in the same DNA strand and of chemical modifications to some of the histone proteins around which double-stranded DNA is wrapped. Some histone modifications (*solid triangles*) loosen DNA's association with the histones and facilitate gene activity, while other modifications (*solid circles*) tighten DNA's wrapping around histones and repress gene activity.

relationships . . . that provide function to the DNA code, which will allow a fuller understanding of normal development, aging, abnormal gene control in cancer and other diseases, as well as the role of the environment in human health" (Jones and Martienssen 2005, 11241). In 2005, the editor-in-chief of the journal *Cancer Research* strongly supported establishing a HEP, writing, "Therapies using the epigenome as a target have the capacity to deliver on the promise of genomic/molecular medicine" (Rauscher 2005, 11229).

Since 2005, impressive advances in relating the epigenome to cancer causes, diagnosis, and therapy presage a revolution in our understanding and treatment of cancers (Esteller 2011). For example, by 2010 approximately three hundred epigenetically modified genes had been associated with cancer cells. Some of these genes are so-called tumor-suppressor genes whose activity normally prevents normal cells from undergoing transformation into cancer cells. An epigenetic modification called methylation can inactivate genes.[13] When a tumor-suppressor gene is methylated, cancer can ensue. Examination of the methylation pattern in a certain gene is now a powerful diagnostic tool for prostate cancer. Finally, the epigenetic modifications on certain genes can indicate the most effective chemotherapy for certain brain tumors and breast cancers. It appears certain that learning more about the epigenome of cancer cells will significantly improve the early detection and treatment of many types of cancer.

Human behavior is also subject to epigenomic changes induced by environmental signals, including a person's experiences (Berreby 2011). For example, evidence indicates that connections exist between childhood abuse, epigenetic changes in the brain, and suicidal behavior. Behavioral epigeneticists also study connections between the epigenome and long- and short-term memory, drug addiction, trauma, child aggression, and depression.

Recently, researchers seeking the genetic basis for multiple sclerosis (MS) sequenced the genomes and examined the epigenomes of immune system cells from two identical twins, one with the MS and the other without the disease. Although the researchers found no differences in the genomes or the epigenomes in the twins' immune systems (Baranzini et al. 2010), future twin studies will examine other cell types, including brain tissue.

Profoundly important is the fact that we inherit epigenetic states from the epigenomes associated with our parents' eggs and sperm. Propensities for obesity, diabetes, and certain cancers may all be passed on to us via the epigenomes of our parents' gametes. In the end, the most important question is, What controls the assembly of specific epigenomes? It is becoming increasingly clear that environmental factors, including our personal lifestyles, have a great influence on our epigenomes and even those of our children and grandchildren.[14] Recent research with plants also shows that the epigenome can spontaneously mutate, similar to the way base sequences of DNA mutate, to potentially alter patterns of gene expression (*Science Daily*

2011).[15] Amazingly, the mutation rate of the epigenome is about one hundred thousand times faster than that of the DNA itself, suggesting that the epigenome may be a major player in biological evolution.

Chimpanzee and Neanderthal Genome Projects

Two more by-products of the HGP are the Chimpanzee Genome Project (CGP) completed in 2005 and a 2010 draft of the Neanderthal genome, the latter based on DNA extracted from 38,000-year-old bones from Croatia (Green et al. 2010). Chimps and Neanderthals are our closest living and extinct relatives, respectively. The lineages of chimps and humans diverged 5 to 7 million years ago, and Neanderthals separated from the line leading to modern humans only about 400,000 years ago (fig. 4.2).[16] Modern humans and Neanderthals cohabited the same territories in the Middle East and Europe for about 10,000 years before the Neanderthals disappeared about 28,000 years ago, and some scientists actually consider humans and Neanderthals to be members of the same species.

By comparing our genome with that of chimps and Neanderthals, we gain insights into ourselves, including our genetic, physiological, behavioral, and psychological roots. We are fortunate that chimpanzees are still with us so that we can correlate genomic data with real-life observations. Results of the CGP showed that chimp and human genomes are 98.77 percent identical. Evolutionary psychologists and neuroscientists hope to gain insight into the genetic basis of what makes us self-conscious, rational, moral, language-using creatures by examining the 1.23 percent difference between our genomes. The chimp genome may also give clues for better fighting diseases like AIDS. The HIV-1 virus that causes AIDS infects only chimps and humans, but viral infection does not progress to AIDS in chimps as it does in humans. Why is this? Detailed information about chimpanzee genes may give an answer.

By 2010, Neanderthal genome researchers had deciphered 60 percent of the Neanderthal genome. From this work, we learned that all modern humans of non-African descent share 1 to 4 percent of their genes with Neanderthals. Interbreeding that led to the shared DNA occurred between

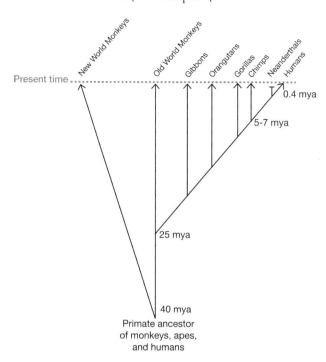

Fig. 4.2. Higher primate ancestral tree showing that chimpanzees are more closely related to humans than to any other living primates, and that the now-extinct Neanderthals co-existed with humans for a while and are very closely related to us. Also shown is the divergence of the hominid (apes and humans) line from Old World monkeys about 25 million years ago (mya).

45,000 and 80,000 years ago, probably in the Middle East, where humans migrating out of Africa first encountered Neanderthals.

Lead author of a Neanderthal genome research article, Richard Green (Green et al. 2010), from the University of California, Santa Cruz, describes the importance of the work, "The Neanderthal genome is a gold mine of information about recent human evolution, and it will be put to use for years to come" (quoted in Ansari, 2010).

Human Genome Project: Human Values

In this section we examine the interface between our increasing genomic knowledge and human values. Specifically, we look at the ethical, legal, and social implications component of the HGP, the responsibility we have as stewards of our epigenomes, and how we are rethinking the morality of using chimpanzees in research.

The Ethical, Legal, and Social Implications Component of the HGP

Before the HGP, government funding for a major science program did not include the stipulation that some of the money be devoted to investigating the project's ethical and social implications. But when the US Congress approved funding for the HGP in 1989, 5 percent of the budget was earmarked for an ethical, legal, and social implications (ELSI) component. Since 1990, theologians, ethicists, scientists, and philosophers in study groups, at conferences, on panels, and as authors of books and articles have fulfilled the ELSI mandate.

Among societal concerns arising from advances in genomic science and addressed by the ELSI scholars are (1) use of genetic information in reproductive decisions; (2) possible stigmatization and negative psychological effects due to genetic differences; (3) designing appropriate policies on patents and commercialization of gene-based products; (4) distinguishing between gene therapy and genetic enhancement and a resurgence of *eugenics*; (5) fairness, privacy, and confidentiality in the handling of genetic information; (6) assessing the role genes play in human behavior and perceptions of genetic determinism; and (7) making informed decisions based on genetic test results with uncertain health implications.

The first four concerns are discussed in other chapters: genomics and reproductive decisions in chapter 3, genetic diversity and stigmatization and patenting of gene-based products in chapter 5, and genetic enhancement and eugenics in chapters 3 and 6. Here we consider genomic information and issues relating to privacy, genetic determinism, and the interpretation and use of genetic tests.

Fairness and privacy. Who should have access to your personal genomic information, and how may it be used? Suppose genomic screening of your DNA at birth reveals a propensity for cardiovascular disease in middle age. This knowledge may motivate you toward healthful diet and exercise habits in your youth in order to minimize your risk, but your risk for developing heart disease may still be higher than that of others your age who are also seeking employment. Should potential employers be able to pass over your job application because you are more likely than your competitors to claim disability wages in the future? Should insurance companies be able to adjust your premiums based on health risks revealed by your genomic information?

In May 2008, President George Bush signed into law the Genetic Information Nondiscrimination Act (GINA), which addresses genome privacy issues. The law protects Americans against discrimination by health insurers and employers based on their genetic information. The bill passed the US Senate unanimously and the House by a vote of 414 to 1 with only Rep. Ron Paul (R-Texas) voting against it.

Genetic determinism. Developmental geneticist Scott Gilbert and his co-authors define genetic determinism as the "view that our genes determine our physical, emotional, and behavioral (observable) traits, irrespective of the environment in which we develop and grow." (2005, 228). Genetic determinism is a tempting view to adopt in this age of modern genomics. Among the thousands of human genes catalogued on the Online Mendelian Inheritance in Man (OMIM) website (http://www.ncbi.nlm.nih.gov /omim/) are ones for dwarfism, deafness, short sleep habit, perfect pitch, epilepsy, early cataract formation, susceptibility to migraine headaches, psoriasis, schizophrenia, many cancers, and age-related macular degeneration. Doesn't it make sense to believe that genes make us who we are and that there is little we can do about it? Well, actually it does not! Biology is replete with examples of environmental factors influencing the expression of genes in plants, animals, and humans.

Sex determination in certain animals is a prime example of the power of the environment over *gene expression.* In some turtles, fish, and snails, sex is determined by environmental factors and not by the animal's genetic makeup. In female honeybees, some become workers and others become a queen depending on the type of food they are fed; plants are famous for taking on different shapes and sizes depending on water and nutrient availability. What about humans? Do we show similar plasticity in response to different environments?

Neuroscientists find that connections between brain cells continually form, disappear, or persist depending upon what we learn and experience. Since behavior and mood depend largely on communications between brain cells, it follows that the environment plays a major role in both. As we saw earlier, our epigenomes control *gene expression.* Studies on cells of identical twins show that as twins age and accumulate different experiences and exposures to different environments, their patterns of gene expression become increasingly different, sometimes resulting in cancer or other gene-based diseases in just one of the twins.[17] These differences likely reflect differences in the twins' epigenomes. In any case, it is clear that persons with virtually identical DNA become unique individuals.[18]

The bottom line is that genetic determinism is a dangerous falsehood. Its destructive consequences include discrimination based on expectations due to genetic pedigree and the belief that one's personal future is completely beyond one's control. Still it is worth noting that one's genetic constitution has been likened to a "genetic leash."[19] Our genes may limit our ability to develop in certain areas or give us propensities to develop in others, but they do not determine the kinds of persons we become. A complex assortment of environmental influences, including friends, education, books read, nutrition, and societal expectations and opportunities, all contribute to who we are.

Interpreting genetic tests. Some of the many genetic tests now available were discussed in chapter 3 in the context of preimplantation and prenatal diagnostics. Most persons in developed countries will soon be able to be tested for genetic diseases and propensities for ill health as adults. Sometimes such tests are useful because they confirm a diagnosis or alert persons to a condition that can be treated. On the other hand, many tests exist for conditions that are untreatable or only warn of an increased risk of acquiring a disease. When healthy, presymtomatic persons undergo such tests, how should positive test results be handled?

Most gene-testing companies do not offer counseling along with the test results, leaving clients on their own to deal with unpleasant results. For example, what would you do if you tested positively for a gene that increases three- to fifteen fold your likelihood of developing Alzheimer's disease. Moreover, there are currently no regulations in the United States to guarantee the accuracy or reliability of genetic testing, some of which is now marketed directly through the mail via do-it-yourself DNA-sampling kits. Authors of the ELSI component of the HGP sum up a section on the pros and cons of genetic testing this way: "Many in the medical establishment feel that uncertainties surrounding test interpretation, the current lack of available medical options for these diseases, the tests' potential for provoking anxiety, and risks for discrimination and social stigmatization could outweigh the benefits of testing."[20]

Taking Care of Your Epigenome

Environments experienced by parents and grandparents may profoundly affect the health of offspring, and these effects may reflect the epigenomes inherited from one's parents (Lalande 1996). Propensities toward diabetes, obesity, and difficulty in managing stress are examples of traits likely to be inherited epigenetically. Even though we do not yet fully understand all the factors that influence our epigenomes, it seems prudent to assume that our lifestyles are somehow reflected in the epigenomes of our children. Physiological effects of drug and alcohol use, high-stress jobs and relationships, and quality of nutrition may all contribute to the epigenomes we bequeath to our children, giving deep ethical significance to lifestyle choices of prospective parents.

Chimpanzees and Ethical Research

In light of the near equivalence of chimps and humans at the DNA level, and chimps' similarity to us physiologically and even socially, it is appropriate to ask ourselves some pointed questions.[21] Should chimps be treated like other research animals such as rabbits and mice, or should they be accorded rights similar to those we give ourselves? If research chimps and human subjects unable to give informed consent are treated similarly, who ought to act on a chimp's behalf by consenting to the research? Should human genes ever be incorporated into the chimp genome, or vice versa? Chimp populations in the wild are endangered, and presently there is a moratorium on using NIH monies to breed chimps. Should other funds be used to breed chimps in order to preserve the species even though chimps bred in captivity cannot be introduced into the wild? If so, for what purpose are we preserving the species? Is it ethical to save chimps simply in order to use them for medical research? As for most ethical problems, these questions have no absolute answers. But struggling with them may give us a clearer view of who we are and how we ought to treat other life forms.

Conclusion

The DNA comprising an individual's chromosomes is the genome of that individual. It also represents the genome of the species. The publically funded Human Genome Project (HGP) became an international effort to determine the 3.1 billion base (A, C, G, and T) sequence for the DNA in the twenty-three different human chromosomes. The final version of the human genome base sequence was completed in 2003, and it is freely and publically available on at least four websites.[22] The base sequence for human DNA by itself does not deliver an understanding about how our genes function or the ability to cure genetic diseases. One goal of the discipline of functional genomics is to use genomic information for improving human health. Major benefits of the HGP are the establishment of an international network of collaborators and research centers for ongoing genomic research and the genomic databases invaluable for applied medical research. Epigenomes are patterns of chemical modifications superimposed over the base sequences of the genomes for humans and other organisms. One's epigenomes vary according to cell type and help regulate the expression of genes in different tissues and disease states. The epigenome is influenced by many environmental factors in complex ways not yet fully understood. A consortium of research institutes recently undertook the Human Epigenome Project to decipher human epigenomes. Human genomic science and technology raise many ethical, legal, and societal issues, including privacy, confidentiality, stigmatization due to genetic differences, reproductive choices based on genetic testing, uncertainties about the meaning of genetic test results, mistaken notions of genetic determinism, the patenting and commercialization of gene-based products, and accessibility to benefits. Recently gained information about the chimpanzee and Neanderthal genomes gives insights into human evolution, human nature, and how to better treat certain diseases.

Questions for Thought and Discussion

1. What other goals/projects would you like to see tackled using the IHGSC as a model for international cooperation?

2. Should companies marketing genetic testing to the public be required to offer professional genetic counseling along with their test results?

3. How much would you really like to know about your personal genome in terms of adult onset diseases or your genetic endowment for certain talents? How much genomic information would you like to know about a potential mate? How much of your genome would you like your fiancé/fiancée to "see"?

4. If you could have a genetic test, at no cost to you, for an above-normal risk of developing Alzheimer's disease, would you have it done? Assume that there is no cure or effective treatment for the disease.

5. Do you believe that most of your family and friends know what the HGP is and what it accomplished? If not, do you think it would be good and useful for them to know these things?

6. What, specifically, do you think is an appropriate level of genetic literacy for citizens in your country? If most citizens are not already genetically literate, what do you think are the best ways to remedy the situation?

7. How should we treat chimpanzees?

8. Do you believe there are major ethical or societal implications of genomic science and technology that were not mentioned in this chapter? What are they?

Notes

1. Specifying the genome derived from eukaryotic chromosomal DNA as the *nuclear genome* distinguishes it from the mitochondrial genome also present in animals and plants and from the chloroplast genome present in green plants. It is called the nuclear genome because chromosomes are located in the cell's nucleus.

2. See the Human Genome Project's website at http://www.ornl.gov/sci/techresources/Human_Genome/home.shtml (accessed May 9, 2012).

3. Ibid.

4. This paper reports on the high-accuracy sequencing of 99 percent of the portion of the genome containing active genes, i.e., 2.85 billion bases. In this classic paper is also reported the surprising finding that the human genome contains only 20,000–25,000 protein-coding genes.

5. The diploid genome consists of both members of each pair of chromosomes, for a total of over 6 billion bases.

6. Collaborating companies with the goal of a $100 genome are BioNanomatrix (now BioNano Genomics http://www.bionanogenomics.com/ [accessed June 20, 2012]) and Complete Genomics (www.completegenomics.com [accessed June 9, 2012]). The work is federally funded by the National Institute of Standards and Technology–Advanced Technology Program (NIST-ATP).

7. In an interview for *Nature* magazine, Michael Egholm, vice president of research and development at the biotech company, 454, stressed what little is known about how to read the meaning of human genomic data. Egholm's comments are quoted in Wadman 2008, 788.

8. A nonmedical benefit of personal genomics for persons interested in family history is the rapid tracing of family origins back tens of thousands of years based on genomic analyses. One provider (www.oxfordancestors.com [accessed June 9, 2012]) claims to identify from which of thirty founding human lineages your family is derived, based on your DNA.

9. Detailed information about the mission, focus area, and details about participating in the project are at http://www.personalgenomics.org (accessed June 9, 2012)

10. The most common DNA modification is methylation of the base C when it occurs just before a G base. Generally, DNA methylation inactivates genes in which it occurs. Histone proteins function to package eukaryotic DNA within chromosomes. Histone modification by methylation, acetylation, or phosphorylation can loosen or tighten DNA packaging and thereby make genes accessible or inaccessible to being expressed. Generally, histone methylation and phosphorylation are associated with inactive DNA, and acetylation with active DNA.

11. The metaphor appeared in the article "Human 'Epigenome' Project" on December 21, 2005, in HUM-MOLGEN news, an international Internet resource on HUMan MOLecular GENetics: http://hum-molgen.org/NewsGen/12-2005/000025.html (accessed August 24, 2009).

12. Organization of the Human Epigenome Project was proposed in 2005. Current members of the consortium are the Wellcome Trust Sanger Institute, Epigenomics AG, and the Centre National de Génotypage. The consortium's website is at http://www.epigenome.org (accessed June 9, 2012).

13. DNA methylation involves replacing a hydrogen (H) atom with a methyl group (CH_3) on several C bases that lie in a region of DNA just in front of the start point for a gene. Methylation of this region of DNA inactivates the nearby gene by preventing it from being read to produce messenger RNA molecules needed for protein production. For a clear, concise, nicely illustrated description of DNA methylation and other epigenetic modifications, see Kubicek 2011.

14. NOVA does an excellent job of explaining the significance of the epigenome to the health of our descendents in its documentary film *Ghost in Your Genes* and through freely available interviews and an interactive video with epigenome researchers at http://www.pbs.org/wgbh/nova/genes/ (accessed May 18, 2011).

15. For a nice summary of the technical report, see Schmitz et al. 2011.

16. The estimated time for divergence of chimp and human lineages is based on fossil and genetic data. The estimate for the Neanderthal-modern human split is based on genomic data analyzed by James Noonan and coworkers (Noonan et al. 2006).

17. These studies were cited in the article "Human 'Epigenome' Project" on December 21, 2005, in HUM-MOLGEN news, an international Internet resource on HUMan MOLecular GENetics: http://hum-molgen.org/NewsGen/12-2005/000025.html (accessed August 24, 2009).

18. Although identical twins are generally assumed to have absolutely identical base sequences in the DNA of all of their chromosome, this may not necessarily be so since recent research shows that the nuclear DNA in identical twins can differ in the number of copies of certain base sequences, a phenomenon associated with disease resistance, autism, and lupus (Bruder et al. 2008).

19. The "genetic leash" metaphor has been used by Harvard University biologist Edward O. Wilson.

20. See http://www.ornl.gov/sci/techresources/Human_Genome/medicine/genetest.shtml#regulate (accessed August 26, 2009).

21. Not only are chimps tool users, but they also display a sense of self and have recently been shown to develop cultures.

22. These websites are the National Center for Biotechnology Information (NCBI), at http://www.ncbi.nlm.nih.gov/projects/genome/guide/human/ (accessed June 9, 2012); the University of California at Santa Cruz, at http://genome.ucsc.edu/cgi-bin/hgGateway (accessed June 9, 2012); the Ensembl Project, at http://uswest.ensembl.org/index.html (accessed June 9, 2012); and the European Molecular Biology Laboratory's (EMBL) Nucleotide Sequence Database, at http://www.ebi.ac.uk/genomes/eukaryota.html (accessed June 9, 2012).

Sources for Additional Information

Begley, S. 2000. "Decoding the Human Body." *Newsweek,* April 10, 50–57.

Brunham, L. R., and M. R. Hayden. 2012. "Whole-Genome Sequencing: The New Standard of Care?" *Science* 336: 1112–13.

Campbell, N. A., J. B. Reece, and L. G. Mitchell. 1999. *Biology.* 5th ed., San Francisco, CA: Pearson-Benjamin Cummings. Information for the primate phylogenetic tree came in part from this source (656–65).

The Chimpanzee Sequencing and Analysis Consortium. 2005. "Initial Sequence of the Chimpanzee Genome and Comparison with the Human Genome." *Nature* 437: 69–87.

Cowley, G., and A. Underwood. 2000. "A Revolution in Medicine." *Newsweek,* April 10, 58–62.

Dreger, A. D. 2000. "Metaphors of Morality in the Human Genome Project." In *Controlling Our Destinies,* edited by P. R. Sloan, 155–84. Notre Dame, IN: University of Notre Dame Press.

Gibbons, A. 2010. "Close Encounters of the Prehistoric Kind." *Science* 328: 680–84. An excellent nontechnical description of how the Neanderthal genome was deciphered and what the results mean for us today.

Gibson, G., and S. V. Muse. 2002. *A Primer of Genome Science.* Sunderland, MA: Sinauer Associates.

Gilbert, S. F., A. L. Tyler, and E. J. Zackin. 2005. *Bioethics and the New Embryology: Springboards for Debate.* Sunderland, MA: Sinauer Associates. See chapter 14, "Genetic Essentialism," 227–40.

Human Genome Project Information. http://www.ornl.gov/sci/techresources /Human_Genome/project/about.shtml (accessed May 9, 2012).

The Internet Encyclopedia of Science. http://www.daviddarling.info/encyclopedia /P/primate.html (accessed May 9, 2012). Information on the primate family tree came in part from this website.

Katsnelson, A. 2010. "Twin Study Surveys Genome for Cause of Multiple Sclerosis." *Nature* 464: 1259.

National Human Genome Research Institute, National Institutes of Health. "The Human Genome Project Completion: Frequently Asked Questions." http:// www.genome.gov/11006943 (accessed August 24, 2009).

Sapolsky, R. 2000. "It's Not 'All in the Genes.'" *Newsweek,* April 10, 68.

Scientific American. 2003. "Celebrating the Genetic Jubilee: A Conversation with James D. Watson." *Scientific American* (April): 66–69.

"Sequencing Whole Genomes: Hierarchical Shotgun Sequencing v. Shotgun Sequencing." http://www.bio.davidson.edu/courses/GENOMICS/method /shotgun.html (accessed July 30, 2009).

Summary of H.R. 493, The Genetic Information Nondiscrimination Act of 2008. "To prohibit discrimination on the basis of genetic information with respect to health insurance and employment." http://thomas.loc.gov/cgi-bin/bdquery /z?d110:HR00493:@@@D&summ2=m& (accessed August 23, 2009).

Thompson, D., F. Golden, and M. D. Lemonick. 2000. "The Race Is Over." *Time,* July 3, 18–23. This article details the meetings between Francis Collins and Craig Venter that led to their agreement to jointly announce completion of a draft sequence for the human genome and their agreeing that a gene patent is appropriate only after the function of the gene is well understood.

Zalmount, I. S. 2010. "New Oligocene Primate from Saudi Arabia and the Divergence of Apes and Old World Monkeys." *Nature* 466: 360–64. This research report puts the divergence of the hominid (apes and humans) line from the Old World monkeys between 28–29 and 24 million years ago.

5
Human Diversity, Genes, and Medicine
Richness and Dangers

We allow our ignorance to prevail upon us and make us think we can survive alone, alone in patches, alone in groups, alone in races, even alone in genders.

—Maya Angelou, Centenary College of Louisiana address

Human Diversity: The Biology

This chapter is about genes, human diversity, scientific correctness, health, and biomedical research. The word *race* is not in the chapter's title because this book is about science and ethics, and the concept of race is both scientifically problematic and ethically treacherous. *Race* is scientifically problematic because it evades a clear, agreed-upon definition and has little to do with biological reality. Using the term is ethically treacherous because it is so value laden. Conversations about race and genetics are easily misinterpreted and misused. Unfortunately, *race* is used frequently in research reports and other writings about human health and diversity, so the term appears within the chapter when we discuss some of those works.

In the first part of the chapter, we will see what scientists have to say about race. Then we look at how scientists study human diversity, what disparities in health and health care exist, how these might be alleviated, and what opportunities the new discipline of pharmacogenomics has for improving human health. In the second part of the chapter, we return briefly to the word *race* used as a proxy for human genetic variation and consider possible dangers lurking in human diversity studies. Specific questions addressed in this chapter include the following:

1. How real is the concept of race?
2. What is the International Human Haplotype Map Project?
3. Are identifiable groups of people more prone to certain diseases or other health problems? If so, why?
4. Do different groups of persons respond differently to certain medicines?
5. What is pharmacogenomics?
6. What ethical pitfalls accompany human diversity studies?

How Real Is "Race"?

We need two new terms for this discussion: *genotype* and *phenotype.* The genotype of an individual, human or non-human, refers to its genetic constitution. On the other hand, phenotype refers to an individual's physical, behavioral, and psychological traits. Hair and eye color, stature, musculature, gregariousness, intelligence, and musical ability are all phenotypic traits. A person's phenotype results from a complex interplay between genotype, the epigenome, and the environment.

Scientists Speak Out on "Race"

Whether *race* is a biologically meaningful term is a scientific question, so we turn to scientists and scientific organizations for an answer. Among the faculty in my Department of Biological Sciences at Auburn University are some experts in categorizing animals, i.e., systematic zoologists, with outstanding national and global reputations. I put the question to two of them, one specializing in invertebrate animals (those with no backbones, from jellyfish and worms to octopuses and sand dollars) and the other in mammals. The invertebrate zoologist replied that race is not a concept applied to the world of invertebrates. The mammalogist said that race is seldom used in the scientific literature; and on those rare occasions when it is used, it refers to a subspecies, a geographical variant within a species. The bottom line is that modern-day biologists get along just fine without using *race* to refer to either genotypic or phenotypic diversity within species of non-human animals. So, biologically speaking, there is no reason to apply the term to *Homo sapiens.*

Next we turn to a highly regarded group of scientists whose profession is to study all aspects of humanity, the American Anthropological Association (AAA), for a scientific opinion on race. In 1998, the executive board of the AAA adopted a statement on race, drafted by a committee representing American anthropologists. The statement is less than three pages long and worth reading in its entirety. Here I quote sections of the statement that describe what race is not and what it is:

> human populations are not unambiguous, clearly demarcated, biologically distinct groups. Evidence from the analysis of genetics (e.g., DNA) indicates that most physical variation, about 94%, lies *within* so-called racial groups. Conventional geographic "racial" groupings differ from one another only in about 6% of their genes. This means that there is greater variation within "racial" groups than between them.
>
> From its inception, this modern concept of "race" was modeled after an ancient theorem of the Great Chain of Being,[1] which posited natural categories on a hierarchy established by God or nature. Thus "race" was a mode of classification linked specifically to peoples in the colonial situation. It subsumed a growing ideology of inequality devised to rationalize European attitudes and treatment of the conquered and enslaved peoples. Proponents of slavery in particular during the 19th century used "race" to justify the retention of slavery. The ideology magnified the differences among Europeans, Africans, and Indians, established a rigid hierarchy of socially exclusive categories, underscored and bolstered unequal rank and status differences, and provided the rationalization that the inequality was natural or God-given. The different physical traits of African-Americans and Indians became markers or symbols of their status differences.
>
> Racial beliefs constitute myths about the diversity in the human species and about the abilities and behavior of people homogenized into "racial" categories. . . . Scientists today find that reliance on such folk beliefs about human differences in research has led to countless errors.

In summary, the AAA statement describes race as a social construct invented in the eighteenth century to facilitate the subjugation of native Americans and persons taken from Africa to be slaves. It notes that human variation occurs in gradations without discrete geographical boundaries. For example, skin color is not just black or white, but occurs in all shades of color that are not restricted to persons indigenous to particular places.

In 2004, just after leading the International Human Genome Sequencing Consortium (IHGSC) to completion of the final draft of the human genome, Francis Collins published an essay titled "What We Do and Don't Know about 'Race,' 'Ethnicity,' Genetics, and Health at the Dawn of the Genome Era." Collins's piece is a summation of a 2003 meeting of sociologists, historians, anthropologists, and geneticists at the National Human Genome Center at Howard University on the topic, called "Human Genome Variation and 'Race.'" Collins acknowledges that *race* and *ethnicity* are terms lacking agreed-upon definitions. He then makes two observations: (1) analyses of genetic variation can fairly accurately predict a person's geographic origin if the person's grandparents all came from the same part of the world, and (2) ancestral origins often have a correlation, although not necessarily precise, with self-identified race. These observations are inconsistent with the claim that race has absolutely no biological correlates. Reflecting on human diversity and disease, Collins observes that the frequencies of occurrence of many genetic variants differ in different regions of the world. Some such variants may be associated with disease.

But genetic factors are not the only cause for health disparities between populations. Collins acknowledges that many health disparities are due to environmental factors such as diet, education, access to health care, stress, social marginalization, discrimination, and such. Collins concludes that a research priority for medical genomics ought to be working to understand the fundamental genetic and environmental causes of health and disease so that we can move beyond using race as a surrogate for disease risk. As director of the National Institutes of Health (NIH), Collins is in a position to institutionalize this priority.

The HapMap Project

The Human Genome Project (HGP) gives information about the number of human genes and their distribution within the twenty-three types of chromosomes, but it alone does not reveal the genetic causes of disease, the chromosomal locations of disease-causing genes, or the frequencies of disease-causing genes among the peoples of the world. To obtain these types of medically important information, a study called the International Haplotype Map (HapMap) Project was undertaken in 2002 and completed in 2005.

The HapMap Project was a collaborative effort by scientists from six countries.[2] Its goal was to "identify and catalog genetic similarities and differences in human beings" so that "researchers will be able to find genes that affect health, disease, and individual responses to medications and environmental factors."[3] Individual human beings differ from each other in just 0.1 percent of the 6.2 billion bases in the DNA of their chromosomes (twenty-three pairs).[4] But 90 percent of this difference is due to substitutions of one of the four bases (A, T, G, and C) by another at particular sites in chromosomes, *single nucleotide polymorphisms* (*SNPs*). The human genome contains about 10 million SNPs scattered throughout it, and the Hap-Map Project has mapped about three million of these in persons with geographically different ancestral origins. Participants donating DNA for the study included ninety persons from Nigeria, ninety from the United States, forty-five from China, and forty-four from Japan.

SNPs come in groups of twenty to seventy that are inherited together as units, each within a stretch of four thousand to sixteen thousand DNA bases called *haploblocks.* The human genome consists of about three hundred thousand haploblocks, which account for 95 percent of the genome.

Each haploblock comes in four to six "flavors," called haplotypes. The specific pattern of bases occurring at the SNP sites along a haploblock characterizes each haplotype. For example, say that in most Chinese persons, a haploblock on human chromosome 3 has the following pattern of SNPs (indicated by letters of specific bases) given below. The periods represent bases that do not vary between individuals, the letters represent bases that do vary between some individuals (SNPs), and the slashes mark the two ends of the haploblock.

/ . . . A . . . . . . C . . T . . . . G . . T . . . /.

This particular pattern of SNPs represents one haplotype of this particular haploblock on chromosome 3. The same haploblock in a person from Japan might look like this:

/ . . . G . . . . . . C . . T . . . . G . . C . . . /.

And the haploblock from most Nigerians might look like this:

/ . . . G . . . . . . T . . C . . . . G . . C . . . /.

Here we have three possible haplotypes for the haploblock on chromosome 3 that we are considering. Notice that all five SNPs occur at the same locations in all three haplotypes but that the identity patterns of the SNPs differ between haplotypes. Every disease-causing mutation is inherited along with the SNPs of the haplotype within which the mutation arose.

This is a wonderful thing for medical researchers. It is as if a disease-causing mutation is a burglar wearing a red cap, green pants, orange shirt, and purple gloves that he can never shed. If you find a person with that combination of clothing, you have probably found the burglar. Likewise, if you find that everybody with a particular disease also carries a particular haplotype, you know that the culprit gene is amongst those SNPs, at a particular region on a particular chromosome. Figure 5.1 illustrates the way SNPs and disease-causing genes hang together within haplotypes during the DNA mixing that accompanies egg and sperm formation.

The HapMap Project mapped the locations of the several hundred thou-

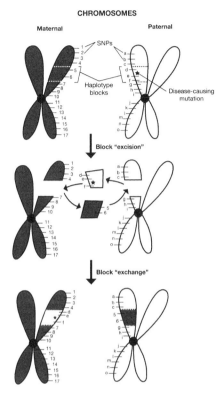

Fig. 5.1. Diagrammatic representation of how single nucleotide polymorphisms (SNPs) and genes with disease-causing mutations are inherited together in haploblocks. During egg and sperm formation (chapter 1; meiosis in fig. 1.10), duplicated maternal and paternal chromosomes exchange corresponding segments of chromosomes (haploblocks) containing genes and SNPs. Haploblocks have variants called haplotypes, each with its own characteristic set of SNPs. Shown here is an exchange between variant haplotype blocks containing SNPs 5, 6 (on maternal chromosome) and SNPs d, e, f, plus a disease-causing mutation (on paternal chromosome). Persons manifesting the disease will all have haplotype block d, e, f. Since the location of the "d, e, f" haplotype block in the genome is known from HapMap Project results, researchers will know where to look for the disease-causing mutation.

sand haplotypes carried by each of the 269 project participants. Analyses showed a relationship between one's genomic pattern of haplotypes and her or his geographic site of ancestral origin. The reason for this is that the SNPs in a haploblock stay together as a unit, even after many generations of genetic mixing.

The HapMap also helps researchers identify the genetic contribution to disorders caused by a complex array of factors including genes, lifestyle, and environment. Such disorders include obesity, diabetes, heart disease, hypertension, many cancers, autism, asthma, dementia, cleft palate, and mental retardation. Finally, information from the HapMap aids discovery of genes affecting responses to drugs, disease susceptibility, and protection from disease. Detailed discussions of scientific and ethical aspects of the HapMap Project are at the NIH's National Human Genome Research Institute website.[5]

The Human Genome Diversity Project and the "1000 Genomes Project"

The Human Genome Diversity Project (HGDP) is organizationally unrelated to either the HGP or the HapMap Project. Initiated in the 1990s, the HGDP aimed to examine the less than 1 percent of the genome that differs from individual to individual in order to gain information about (1) the origins of individual "racial" groups, (2) the genetic bases for human traits and how they relate to racial and small, native populations, and (3) the genetic causes for "race"-based vulnerabilities to certain diseases. The HGDP described its target populations in geographical, ethnic, and linguistic terms. DNA samples were obtained from a preexisting bank of cell lines collected from worldwide donors and maintained at a French institute for human-population genetic studies.

The project was severely criticized for using socially constructed categories to guide its DNA sampling. Concerns were also raised that the data could be used to exploit or harm indigenous populations (South and Meso American Indian Rights Center 1995). Although the HGDP no longer operates as the original international consortium, data gathered from 927 individuals from fifty-two populations was published in 2006 (Conrad et al. 2006). Overall, the HGDP data confirmed the medical usefulness of haplotype maps obtained by the less controversial HapMap Project that sampled *Homo sapien* DNA based on geographical ancestral home.

In 2008 a new project to examine the genetic diversity in *Homo sapiens* was initiated, The 1000 Genomes Project. The goal of the project is to identify the genetic variants that occur at a frequency of 1 percent or higher in our species. By project's end, DNA samples from twenty-five hundred individuals from twenty-eight different populations, representing five of the world's major human ancestral regions (Europe, East Asia, South Asia, West Africa, and the Americas), will have been sequenced. Scientists and genomic research institutes in the United Kingdom, United States, and

China are collaborators in this task. The project's name comes from the goal of a pilot study completed in June 2009: sequencing to a high level of accuracy and completeness one thousand genes from one thousand individuals.

Biomedical researchers will use data from the 1000 Genomes Project in conjunction with information provided by the HapMap Project. By describing most of the DNA variants that exist in our species, the 1000 Genomes Project is like the zoom function on a camera for locating sites in the human genome associated with diseases and other traits such as disease resistance and drug metabolism. If the human genome is a forest and a disease-causing variant is a blackbird sitting on a very small branch of a particular tree, the HapMap can tell which tree (haplotype) the bird is in, and 1000 Genomes Project data can home in on the tiny branch (gene segment) in that tree where the bird is perched. Project data will be used not only to learn about diseases caused by variants of single genes, such as cystic fibrosis and Huntington disease, but also to learn about diseases with multiple, causative, genetic factors, such as heart disease, many cancers, and others listed earlier.

Biogeography, Human Diversity, and Human Health

In this section we see how certain human traits are evolutionary adaptations to particular climates and other geographical and environmental factors. Then we examine relationships between human diversity and health, and the issue of group medicines.

Early human migrations and human diversity. Modern genetics and paleontology (the study of fossil remains) tell us when and where on the planet our species originated and about our ancestors' early migrations. Analyzing the sequence of bases in mitochondrial DNA is especially useful for human ancestry studies because mitochondria are inherited only from mothers, whereas nuclear, chromosomal DNA contains a complex mixture of genes from both maternal and paternal ancestors all the way back to the beginning of the species. By examining mutations in mitochondrial DNA from indigenous populations worldwide and knowing the rate at which mutations accumulate in mitochondrial DNA, our origins and early migratory patterns can be deduced.

Molecular geneticists and paleontologists agree that *Homo sapiens* arose in Africa, probably near the Rift Valley near modern-day Kenya, about 200,000 years ago. Human populations in Africa were relatively small and isolated from each other for at least 100,000 years. Then about 60,000 years ago, we began leaving Africa and eventually populated every continent except Antarctica. We were living in Indonesia and Australia over 40,000 years ago, northern Europe by about 35,000 years ago, eastern Asian by 25,000 years ago, and moving into the Americas 15,000–35,000 years ago (fig. 5.2). Humans living in different regions of the world faced different environments and experienced correspondingly different evolutionary selective pressures. Individuals with adaptive traits tended to have more offspring who lived to become parents themselves. Gradually, some regional physical and physiological differences appeared. Examples are body type, skin color, nose shape, the sickle-cell trait, and the ability to digest cow's milk.

In general, the closer a person's ancestors lived to the equator, the darker is his or her skin. Dark skin is due to melanin, a skin pigment that protects the skin against DNA damage due to the sun's ultraviolet (UV) light. The intensity of UV light is greatest at the equator and lessens as one moves north or south from the equator. So, skin coloration is a beneficial adaptation for living in tropical and sub-tropical regions. But UV light also has an important benefit, namely, to initiate vitamin D formation in skin cells. Vitamin D is essential for absorption of dietary calcium needed for healthy bones, teeth, and immune systems. As one moves further away from the equator, less UV light is available, so *natural selection* produces lighter skin in order to facilitate vitamin D production, preventing vitamin D deficiency and diseases like rickets. In tropical regions, sunlight is so intense that persons produce enough vitamin D even with darker skin.

The intensity of skin pigmentation in humans results from the combined action of several different genes. In 2005, researchers in Pennsylvania, Ohio, Texas, Utah, Indiana, and Canada collaborated to identify a

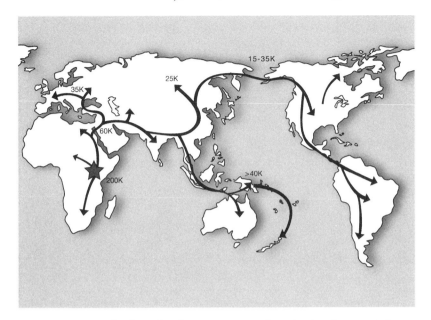

Fig. 5.2. Summary of early modern human migrations from the site of our species' origin (*star*) in East Africa about 200,000 years ago (K, kilo-years, thousands of years). Figure adapted from Cavalli-Sforza, Menozzi, and Piazza 1994 and Hirst 2009.

gene on human chromosome 15 that is responsible for 25 to 38 percent of the difference in skin color between native Africans and Europeans (Lamason et al. 2005). The gene was discovered by first studying the genetic basis for pigmentation in a fish species. The structure of the human gene is very similar to that of the fish gene, indicating that the fish gene is ancestral to the human gene. In fact, the human gene product increases the level of fish pigmentation when transferred into lightly pigmented fish. Examination of human genome databases from the HapMap Project shows that in humans the African and East Asian variant is ancestral to the European variant of the gene.

Body shape is also a product of natural selection. Persons indigenous to tropical and subtropical regions tend to have longer limbs and thinner bodies than do persons with ancestors from colder climates, who tend to have shorter limbs and rounder bodies. We lose body heat through our skin. The longer and lankier we are, the more surface area we have compared to our body's volume, and the more rapidly we can cool ourselves. Long, lanky limbs behave like a car's radiator. Conversely, the rounder and shorter we are, the better we are at conserving body heat. Similarly, a long, narrow nose is an adaptation to colder climates, being better at warming and moistening air before it reaches the lungs. Lower, wider noses are more frequent among persons indigenous to hot, humid climates.

The sickle-cell trait occurs in persons whose ancestors lived in regions harboring malaria-carrying mosquitoes. Having a single copy of the gene for sickle-shaped red blood cells lends resistance to malaria.[6] The ability to digest milk sugar, lactose, as an adult is a trait that was selected for in populations that received a major portion of their dietary protein from animal milk, cheese, and other dairy products, especially Finns, Swiss, and Swedes.

Every human population is a kaleidoscope of genotypes since variants are arising all the time and there are no clear borders defining environmental regions or breeding populations. Therefore, no physical or physiological trait is strictly limited to a specific population. Moreover, none of the adaptive traits just discussed, including skin color, is a valid basis for using a subspecies (racial) scheme for classifying modern humans.

Human diversity: health and drug responses. Dramatically higher frequencies of heart attacks, diabetes, cancers, low birth weight, tuberculosis, and HIV/AIDS occur among African Americans than among European Americans (Agency for Healthcare Research and Quality 2008; Krieger et al. 2005). How much these and similar health disparities are due to genetic factors is not known. However, it is virtually certain that non-genetic factors such as access to and use of high-quality health care, social and economic environments, and cultural practices including diet are major causes of health disparities like those just named. In the case of cancer, underrepresentation in clinical trials may also aggravate health inequality. Women, the elderly, and ethnic minorities are underrepresented in clinical trials for cancer, compared to the incidence of cancer in these groups (Murthy, Krumholz, and Gross 2004; Agency for Healthcare Research and Quality 2000). Identifying and working to eliminate non-genetic causes for health disparities between groups are imperatives for a just society, but this topic is outside the focus of this book. Therefore, let's turn to genetic factors known to contribute to health disparities and how biotechnology can help to improve human health worldwide.

The incidence of carriers for some genetic diseases is much higher among certain groups of people (American-Israeli Cooperative Enterprise 2012; Goljan 2004). A well-known example is the Ashkenazi Jewish population of Eastern European descent. Twenty-five percent are carriers for diseases including Tay-Sachs, Fanconi anemia, and cystic fibrosis. Thalassemia, a blood disease caused by abnormal hemoglobin molecules, is another example. The disease has a markedly higher incidence in persons with ancestry from the Mediterranean, North Africa, and South Asia.

As for responses to specific drugs, there is a vast medical literature that reports "racial" and ethnic differences.[7] For example, the drug Bi-Dil (a vasodilator) is reported to be more effective in African Americans than in persons of European ancestry;[8] propanolol (a non-selective beta-adrenoceptor blocker) is reported to be more effective in Europeans than in African Americans for initial treatment of hypertension; and haloperidol (an antipsychotic) is reported to produce a higher frequency of side effects in Asians than in Europeans. As interesting as reports like these are, their value is often limited when it comes to prescribing medicine for a particular individual. First, all members of an identified group do not respond in the same way. Second, just because a difference between groups is observed does not mean that the difference has a genetic component. Finally, as we will see later for the case of BiDil, some studies may be biased or improperly designed.

There is at least one notable exception to the frequent ambiguity of genetic causes for apparent links between ethnicity and response to drugs. For carbamazepine, a drug used to treat epilepsy and diverse mental disorders, a high risk of life-threatening side effects in persons carrying a gene called HLA-B*1502 is well established (Lim, Kwan, and Tan 2008). The frequency of the HLA-B*1502 gene is highest in Han Chinese, Malay, and Thai persons. So these ethnic groups, and possibly other Asian peoples for which the prevalence of HLA-B*1502 is not yet known, are especially vulnerable to severe reactions to carbamazepine. A simple genetic test can determine whether a person carries the HLA-B*1502 gene.

To summarize, it is apparent that human health could be enormously improved by better health habits and expanded access to high-quality health care, including regular exams, screening and diagnostic procedures, and disease treatment. However, causes for some diseases, and probably the efficacy of some medicines, have strong genetic components, and these genetic components often distribute themselves unequally across the spectrum of human diversity.

Pharmacogenomics

A 2006 report listed adverse drug reactions as the fourth leading cause of death in the United States after cancer, heart disease, and stroke. The branch of pharmacology called pharmacogenomics takes genomic information about individuals or groups of persons and uses it to design more personalized and safer drug treatments. One advantage of personalized, genomic medicine is its potential to reduce the risk of adverse drug reactions and optimize drug therapies by tailoring diagnostic protocols, drug treatments, and preventative medicine to the genetic makeup of the patient.

Diagnosis, therapy, and prognosis for breast cancers epitomize an especially fruitful collaboration between genomics, pharmaceutics, and the clinic. Breast cancer is not just one disease. There are many subtypes and risk groups of breast cancers that manifest themselves in differing *gene expression patterns* in tumor cells (Foekens et al. 2008). To establish which of the twenty thousand to twenty-five thousand genes in the human genome are active in a tumor cell biopsy, clinicians use a *gene expression microarray*. This powerful laboratory technology makes use of the fact that an RNA molecule (or its complementary DNA) will bind specifically to DNA fragments with base sequences like those in the gene from which the RNA was transcribed.[9] To perform a gene expression microarray test, technicians first attach thousands of microscopic samples of DNA, each representing a different gene, to a glass surface. The DNA-spotted surface is then exposed to fluorescently labeled molecules representing all of the RNAs produced by the biopsied cells. A computerized analysis of the intensity and color of the resulting fluorescence at each spot reveals the level of activity of each gene in the examined cells; that is, their gene expression pattern.

Results of gene expression microarray tests have practical applications. For example, microarray results can identify "poor prognosis" and "good prognosis" cancers. For an early-detected "poor prognosis" cancer, aggressive chemotherapy may be recommended to prevent the cancer from spreading to other organs. On the other hand, a woman with a "good prognosis" cancer, one known not to metastasize, may be spared chemotherapy even after a relatively late diagnosis.

In 2010, researchers in Norway reported using microarray analysis of blood cells for highly accurate and very early detection of breast cancer (Aaroe et al. 2010), and other workers at the University of Texas demonstrated the value of microarray analysis for choosing the best chemotherapy strategy for certain breast cancers (Symmans et al. 2010).

Another dimension of pharmacogenomics is rational drug design, the development of a drug based on detailed molecular knowledge about the drug's target. Drug targets are usually proteins that directly or indirectly cause disease such as cancers. The strategy is to design a drug molecule that interacts specifically with the target to neutralize its harmful effects. A frequently encountered problem is that a drug's target protein(s) may be essential for healthy cell function as well as for cancer cell function.

In the case of chronic myelogenous leukemia (CML), combined cell biology, genetic, genomic, and *proteomic research* led to the discovery that a single abnormal protein is responsible for the disease. Rational drug design work then produced the drug imatinib (FDA approved in 2001 and marketed by Novartis as Gleevec). Imatinib is the first anticancer drug that acts to specifically inhibit the action of a single protein known to cause a particular cancer. It contrasts with other chemotherapeutic drugs that attack all rapidly dividing cells, including healthy cells needed for blood formation and hair growth.

Although pharmacogenomics will undoubtedly make drug treatments safer and more effective for many persons, it is not a panacea for all drug treatment risks. For example, even when genes affecting drug efficacy are identified, they may be expressed differently in different individuals or groups due to differences in underlying physiologies or genetic background. So simply testing for the presence of a genetic variant will not give foolproof information for designing drug therapies with no adverse side effects. Finally, designing new drug trials based on genetic screens predicting drug efficacy could lead to exclusion from the trials of persons with genetic profiles unfavorable for the trials. Such exclusions could lead to "orphan genotypes" left unresearched or untreated, even though persons with these genotypes might respond favorably to the drug.

Human Diversity: Human Values

The greatest ethical issue lurking around genomic studies of human diversity is the danger of giving scientific credence, or the perception of reality, to the notion of human races, i.e, reifying *race*. The HapMap Project tried to guard against this by collecting DNA from donors selected on the basis of "continent of origin," those having at least four grandparents who were born in the specified geographical region. The problem does not go away though, because geographical groupings tend to follow the socially constructed racial groupings.

If we wish to use genomic information to help improve health and health care for all of the world's people and we also wish to rid ourselves of destructive racial thinking, we have a dilemma. Despite fuzzy genetic boundaries between geographical groupings of people, different groups of humans do show differences in the frequencies of genes for some diseases, drug responses, and other medically relevant traits. Therefore, genomic studies that generalize their results to groups can sometimes be medically beneficial. Should we risk reifying race in order to obtain these benefits?

If we could cheaply and rapidly examine the genome of every patient for genes germane to her or his condition, there would be no point in using race as a proxy for personalized genomic information, and the dilemma would disappear. But, we are caught in a period when it is possible to gather and analyze genomic data from groups of people but not yet feasible to examine the genome of every patient. Perhaps the best we can do is to proceed vigilantly and sensitively with genomic medicine, working toward the day of truly personalized medicine. In the meantime, researchers, volunteer research project participants, community leaders, and the general public need to know about ethical problems associated with gathering and using genomic data in studies like the HapMap, Human Genome Diversity, and 1000 Genomes Projects.

Ethical Issues in Designing and Gathering Data for Genomic Studies Comparing "Racially" or Ethnically Defined Groups

1. *Issue for the researcher*: leaving some persons out of the study because their group size is too small to be economically profitable to pharmaceutical companies or for other reasons. For example, the HapMap Project was criticized for having no Native Americans or Pacific Islanders in its study. But, the decision to exclude Native Americans came after consultations with representatives of the Native American health research community, who voiced concerns about the data being used to facilitate population history studies. Most Native American traditions include accounts of their origins in America near where their recent ancestors lived. Concerns were that genetic data indicating different origins, such as Asia, for Native Americans could be disruptive to their traditions.

2. *Issue for the researcher*: how to explain the study and its potential uses so that potential participants can give their informed consent. This is an especially sensitive issue for African Americans due to the Tuskegee Syphilis Study performed by the US Public Health Service from 1932 to 1972. The study was to learn about the progress of untreated syphilis in African Americans. After penicillin was found to cure syphilis in the 1940s, study participants were not offered the drug nor were they informed of the new cure for their disease. Instead, the study continued for another twenty-five years, resulting in the deaths of several participants and their spouses. Exemplifying how project participants ought to be prepared to give informed consent, the HapMap Project held community engagement events for potential participants before seeking their informed consent. Activities included individual surveys to learn about concerns, focus group discussions, community meetings, and public surveys.

3. *Issue for the participant*: not knowing what the data might show or how it will be used. The participant could be putting his entire group at risk for future stigmatization or discrimination. In the case of the HapMap Project, results are made public for any researcher to use. The hope is that the uses will lead to improved health. But data showing that certain groups have higher risks for certain diseases or other health problems could lead to discrimination. There is also the possibility that unrestricted genetic diversity data could be used to develop biological weapons of disease targeted at particular racial or ethnic groups.

4. *Issue for community leaders helping to recruit participants*: community leaders could unwillingly and unknowingly put their groups at risk due to misuse of the data.

5. *Issue for pharmaceutical companies seeking drug patents and government agencies approving drugs*: reifying race by seeking patents and approving drugs for race-specific use when trials to demonstrate race specificity of drug efficacy are lacking but the prospect of large profits is great. An example is the seeking and obtaining a race-specific patent on the drug BiDil by NitroMed and the 2005 Food and Drug Administration (FDA) approval of BiDil as a race-specific drug to treat heart failure in African Americans. How NitroMed obtained its patent and FDA approval is a complicated and convoluted story.[10] In the end, patent laws and profit motive trumped trials that may have shown BiDil to be effective for non–African Americans as well. In this case, both patenting law and the unique history of BiDil conspired to needlessly reify race and make the drug nearly ten times as expensive as generic equivalents, whose makers are prevented from advertising until 2020.

Ethical Issues Associated with the Use of Results from Genomic Studies Comparing "Racially" or Ethnically Defined Groups

1. *Stigmatization of members in a group due to misuse or misunderstanding of the data*: If a condition is found to have a higher incidence in one group than in others, the mistaken belief may arise that all members of the group have that condition and that the condition does not exist outside the group. This

mistake could encourage racism by fostering the notion that humanity is divided into distinct types, with clearly demarcated boundaries. The Genetic Information Nondiscrimination Act of 2008 (chapter 4) took a big step in guarding against discrimination based on genotype in the United States.

2. *Marginalization of small groups due to business economics*: New drug research and development is expensive, and pharmaceutical companies are in fierce competition with each other. The most profitable drugs are those needed and used by the most people. Finding that certain adverse health conditions exist mainly in small minority groups could move attention away from research on treatments for those conditions.

3. *Confusion of group membership*: Genetic findings could conflict with traditional social or cultural identifiers that groups use to define group membership. Similarly, genetic data could undermine religious or cultural traditions or political status by revealing new information about group ancestry.

4. *Disparities in the ability to use project data for health benefits*: All groups represented in human diversity studies may not be equally positioned to take advantage of the findings to improve group health. This is an issue of distributive justice. It is important to develop strategies for data use so that benefits do not accrue mainly to persons in the wealthiest or most technologically advanced groups.

The overarching question for human genetic diversity studies is how to balance the several risks inherent in genomic research against its great potential to improve the health and well-being of all people.

Conclusion

Anthropologists, social scientists, and biologists widely agree that the traditional use of *race* is biologically meaningless. There are no subspecies of *Homo sapiens.* Modern genomic methods and studies show that humans are overwhelmingly alike. Yet sparse genetic differences between groups with different geographical sites of origin may harbor propensities for adverse or beneficial drug responses, health ailments, or protection against certain diseases. This is not surprising given the history of our species. Migrations from our founding population in East Africa tens of thousands of years ago took us into other continents with vastly different climates, exposures to physical factors like UV light, and disease vectors like malaria-carrying mosquitoes. In response, natural selection produced different frequencies of gene variants in different populations. The International HapMap Project results completed in 2005 provide tools for researchers to identify medically relevant gene variants, develop genetic screens for these, and apply genomic information to improving the health of human beings everywhere. But dangers lurk in the possible misuse and misinterpretation of genomic data. These include stigmatization and discrimination against genetically related groups of people and possible undermining of certain cultural and religious traditions. The greatest danger of genomic studies of human diversity is inadvertent encouragement of racial thinking that has its origins in social oppression, not in benefiting humanity.

Questions for Thought and Discussion

1. How real do you believe *race* is?
2. Does knowing that all modern-day humans arose from a relatively small population of individuals living in East Africa change the way you view yourself or others?
3. How can the HapMap be used to benefit the health of all humans?
4. How might information from the HapMap Project be used to reinforce destructive racial thinking?
5. Can you suggest a term other than *race* to describe groups of persons sharing a common genetic heritage and geographic region of ancestry?
6. Can you think of other ethical issues not mentioned that might arise from genomic studies of human diversity?
7. What are the biggest potential benefit and greatest risk of genomic human diversity studies for you personally?

Notes

1. An authoritative and definitive book about the "Great Chain of Being" concept is Arthur O. Lovejoy's 1936 work, *The Great Chain of Being: A Study of*

the History of an Idea, which had its twenty-second printing in 2001 and was re-copyrighted by the President and Fellows of Harvard College in 1964.

2. HapMap Project collaborators included scientists and funding agencies from Japan, Nigeria, the United Kingdom, China, Canada, and the United States.

3. See the International HapMap Project website, "About the HapMap," http://www.hapmap.org/thehapmap.html.en (accessed September 9, 2009).

4. Recall that human diploid cells have forty-six chromosomes and eggs and sperm have just twenty-three chromosomes.

5. See http://www.genome.gov/11511175#al-4 (accessed May 18, 2011); at this site are described the rationale, scientific approach; ethical and societal issues in DNA sampling; and the policies concerning data access and intellectual property for the HapMap Project.

6. Two copies of the recessive gene for sickle-shaped red blood cells cause sickle-cell anemia, a condition occurring when one receives the genetic trait for sickling from both parents.

7. A table of twenty-eight drugs reported to induce different responses in different racial or ethnic groups, along with the sources for the data, sample size, and descriptions of the responses appears in Tate and Goldstein 2004.

8. Although this is claimed, the primary study resulting in BiDil's patenting and approval as a race-specific drug included one thousand African Americans but no others. Please see Sankar and Kahn (2005) for the complete story.

9. Please see chapter 1 and the glossary for a description of transcription and the relationship between DNA, RNA, and proteins.

10. The story is told well by Sankar and Kahn 2005.

Sources for Additional Information

American Anthropological Association. 1998. "American Anthropological Association Statement on 'Race,'" May 17. http://www.aaanet.org/stmts/racepp .htm (accessed September 4, 2009).

———. 2012. "About Race: A Public Education Project." http://www.aaanet .org/resources/A-Public-Education-Program.cfm (accessed January 20, 2012).

American Anthropological Society of America. 2009. "Race: Are We So Different? A Project of the American Anthropological Society of America." http://www.understandingrace.org/home.html (accessed October 30, 2009).

Campbell, M. C., and S. A. Tishkoff. 2008. "African Genetic Diversity: Implications for Human Demographic History, Modern Human Origins, and Complex Disease Mapping." *Annual Review of Genomics and Human Genetics* 9: 403–33.

Cavalli-Sforza, L. L., P. Menozzi, and A. Piazza. 1994. *The History and Geography of Human Genes.* Princeton, NJ: Princeton University Press.

Collins, F. 2004. "What We Do and Don't Know about 'Race,' 'Ethnicity,' Genetics, and Health at the Dawn of the Genome Era." *Nature Genetics Suppl.* 36 (11): S13–S15.

Goldstein, D. B., and J. N. Hirschhorn. 2004. "In Genetic Control of Disease, Does 'Race' Matter?" *Nature Genetics* 36: 1243–44.

Hirst, K. K. 2009. "Human Migration Map." http://archaeology.about.com/od/stoneage/ss/tishkoff_2.htm (accessed September 6, 2009).

International HapMap Project. 2012. http://www.hapmap.org/index.html.en (accessed May 16, 2012).

Juengst, E. T. 2007. "Population Genetic Research and Screening: Conceptual and Ethical Issues." In *The Oxford Handbook of Bioethics,* edited by B. Steinbock, 471–90. New York: Oxford University Press.

Kassim, H. 2002. "Race, Genetics, and Human Difference." In *A Companion to Genethics,* edited by J. Burley and J. Harris, 302–16. Malden, MA: Blackwell Publishers.

Lee, S. S. 2003. "Race, Distributive Justice, and the Promise of Pharmacogenomics." *American Journal of Pharmacogenomics* 3: 385–92.

Mukhopadhyay, C. C., R. Henze, and Y. T. Moses. 2007. *How Real Is Race? A Sourcebook on Race, Culture, and Biology.* Lanham, MD: Rowman & Littlefield Education.

National Coalition for Health Professional Education in Genetics. 2012. "Race and Genetic FAQ." http://www.nchpeg.org/index.php?option=com_content&view=article&id=142&Itemid=64 (accessed January 18, 2012).

Oubré, Alondra. 2011. *Race, Genes, and Ability: Rethinking Ethnic Differences.* Vols. 1 and 2. Woodland Hills, CA: BTI Press.

Sankar, P., and J. Kahn. 2005. "BiDil: Race Medicine or Race Marketing?" Health Affairs—Web Exclusive, Project HOPE—People-to-People Health Foundation, Inc. http://content.healthaffairs.org/cgi/reprint/hlthaff.w5.455v1.pdf (accessed September 10, 2009).

Tate, S. K., and D. B. Goldstein. 2004. "Will Tomorrow's Medicines Work for Everyone?" *Nature Genetics Suppl.* 36 (11): S34–S42.

Weigmann, K. 2006. "Racial Medicine: Here to Stay?" *EMBO Reports* 7 (3): 246–49.

6

Genetic Enhancement

Humankind Healing and Redesigning Itself

> We embrace technologies that tame and harness nature because we think they improve our lives, and we will accept or reject human genetic manipulation on the same grounds.
>
> —Gregory Stock, *Redesigning Humans: Our Inevitable Genetic Future*

Michelangelo did a wonderful thing when he sculpted the *David* (fig. 6.1). This single creation embodies the confidence, courage, and creative potential of Renaissance Florence and of humankind today. Michelangelo's chisels are now *genes,* and humans are simultaneously the artist and the block of stone. In the decades ahead, we will override *natural selection* and sculpt our genomes. According to what? Our own best judgment? What gives us pleasure? What sells best?

In this chapter we explore how humankind can direct its future evolution and what choices, promises, and perils lie ahead. These questions guide our explorations:

1. What is *genetic engineering?*
2. What are the different types of human genetic engineering?
3. What human genetic engineering is already performed?
4. How might humans genetically alter themselves in the future?
5. What does the public think about genetically engineering humans?
6. Is human genetic engineering safe?
7. At what point does gene therapy become genetic enhancement?
8. Is it morally wrong to tamper with the human genome?

Fig. 6.1. Michelangelo's *David,* Florence, Italy. Photograph by James T. Bradley, 1983.

9. Should parents genetically alter their children without the children's consent?

10. Will human genetic engineering lead to a new *eugenics?*

Human Genetic Engineering: The Biology

What Is Genetic Engineering?

Genetic engineering is the intentional and directed altering of an organism's genome, usually involving just one or a few genes. In plants, this includes putting a daffodil gene, which confers resistance to a chemical herbicide, into soybean plants so farmers can spray to kill weeds without harming the bean plants. In humans, it can mean inserting a normal gene for an important enzyme into the DNA of white blood cells to restore normal function to a patient's immune system. In the future, it might mean inserting genes into eggs or early embryos to increase the cognitive abilities of upcoming generations.

This chapter is about genetically engineering human cells. In it we consider some examples of human cell genetic engineering already performed, some successful and some tragically unsuccessful. Then we look at the prospects for human genetic engineering in the future. Finally, we explore moral issues that come with various types of human genetic engineering.

What Are the Types of Genetic Engineering?

All human genetic engineering involves manipulating DNA in living, human cells. Depending upon the kind of cell being altered and the motive for doing so, there are four types of human genetic engineering. Since human motives are often difficult to define, the boundaries between some types of genetic engineering are also elusive. Nevertheless, categorizing genetic engineering into four types helps us to think about the various ethical issues arising from the technology.

One way to classify genetic engineering is by the cell type being engineered: *somatic* (body) *cells* versus *germ cells* (eggs and sperm). Somatic cell genetic alterations last only for the lifetime of the person receiving the genetically engineered cells; the alterations are not passed on to future generations. By contrast, germ-line genetic alterations are potentially immortal since they can be passed on to future generations.

A second classification of genetic engineering is based on motive: therapeutic versus enhancing. Gene therapy aims to cure a genetic disease or treat some other genetically based health problem. Genetic enhancement, not yet performed in humans, would aspire for cosmetic changes in non-health-related traits. For example, increasing the stature of or diminishing the fat deposits in a normal person would be enhancements.

In their 1997 book, *Ethics of Human Gene Therapy,* bioethicist LeRoy Wal-

Table 6.1. Types of human genetic engineering

Engineered cell type	Somatic	Germ line
Therapeutic	Type 1	Type 2
Enhancing	Type 3	Type 4

Note: There are two classification schemes for human genetic engineering: gene therapy versus genetic enhancement and somatic cell engineering versus germ-line engineering. Together, the two classification methods yield four types of human genetic engineering, each with unique biological characteristics and ethical implications.

ters and molecular biologist Julie Palmer use the following matrix to illustrate how combining the above categories yields four types of human genetic engineering: (1) somatic cell therapy, (2) germ-line therapy, (3) somatic cell enhancement, and (4) germ-line enhancement (table 6.1). Considering how each of these is performed and when they might be used will help us think about the moral issues that they raise.

Somatic Cell Gene Therapy (Type 1)

Somatic cell gene therapy (type 1) is the only kind of genetic engineering currently done in humans. The first clinical trial for somatic cell gene therapy in humans was in 1990, but the technology is not yet routine for treating or curing any disease or other medical condition. Still, recent advances in gene therapy technology and many ongoing clinical trials predict a promising future for type 1 genetic engineering in humans.

Somatic cell gene therapy aims to cure genetic disease by correcting abnormal genes. Ways to do this include (1) inserting a normal gene into the genome in hopes that it will be active and compensate for the nonfunctional gene, (2) replacing the abnormal gene with a normal copy of the gene by an "add and delete" process, (3) repairing the defective gene while it is still in the genome, and (4) altering the abnormal gene, making it more or less active. The first method is the most common approach.

How are somatic cells genetically engineered? The overall strategy for somatic cell gene therapy is to remove some cells from the patient, add normal copies of the malfunctioning gene (along with its adjacent regulatory DNA) to these cells, and then reintroduce the cells into the patient (fig. 6.2). Recall that somatic cells are body cells, as opposed to germ cells (eggs or sperm). Genetic alterations to somatic cells are not passed on to future generations, and the alterations disappear when the person dies.

Somatic cells chosen to be genetically altered are often adult stem cells that will give rise to normal versions of the malfunctioning cells after being reintroduced into the patient. For example, engineering bone marrow stem cells is a suitable approach for treating leukemias and other genetically caused blood diseases. The most common way of inserting new genes into living cells is to use an inactivated virus as a carrier. Such viruses are called *vectors*. Scientists and biotechnicians select an appropriate vector, insert the normal human gene into vector DNA, and then expose the patient's cultured cells to thousands of vector particles carrying the normal gene. The vectors enter the cells, and some of the normal genes they carry become permanently incorporated into the DNA of the host cells.

A viral vector is not the only way to deliver new DNA into a cell's genome. Large quantities of a normal gene can also be injected directly into the affected tissue. By a poorly understood process, some of the injected genes traverse cell membranes and enter cell nuclei, where they become active. Genes may also be delivered to cells inside membrane-bound vesicles called *liposomes* (fig 6.2). When a liposome fuses with the outer membrane of a cell, it empties its genetic cargo into the cell's cytoplasm. Some of the genes then find their way into the nucleus and become active.

Adding normal genes to cells to compensate for abnormal genes is not the only approach to somatic cell gene therapy. Abnormal genes may also be deleted and then replaced or repaired, much more difficult procedures than simply adding a gene at an unspecified location in the genome. Gene replacement or repair has not yet been done in humans. But in 2011 a new genetic engineering technology, zinc finger-mediated gene therapy, successfully inactivated a specific gene in immune system cells of HIV patients. Clinical trial results demonstrated the technology's potential for treating AIDS and curing certain genetic diseases (Sharp 2011). Finally, another technology functionally equivalent to gene deletion, called *RNA in-*

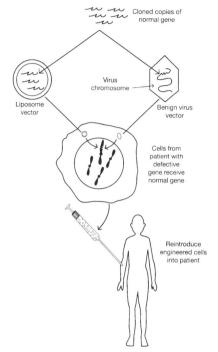

Fig. 6.2. Somatic cell gene therapy. Copies of a normal gene are introduced into cells from a patient with a defective gene. Either a liposome vector or an inactivated virus vector delivers the normal genes into the cells. One or more copies of the normal gene insert themselves into the chromosomal DNA of the patient's cells. Engineered cells reintroduced into the patient function normally because the normal gene compensates for the defective gene.

terference (RNAi), shows great promise as a way to prevent the expression of disease-causing genes.

Clinical trials using somatic cell gene therapy. In 1990 the first somatic cell gene therapy trial successfully treated a four-year-old girl born with a defective gene for an important *enzyme.* The child's T cells, a type of white blood cell critical for a normally functioning immune system, were dying due to lack of an enzyme called ADA needed to rid cells of a toxic by-product of normal metabolism.[1] Victims of ADA deficiency have a life-threatening immunodeficiency that makes them acutely susceptible to infectious disease.

The child was treated by removing ADA-deficient T cells from her body, introducing multiple copies of the normal gene for ADA into the cells, and then injecting the genetically engineered cells back into her body. She received six treatments over a period of several months. The happy result was a much stronger immune system for the child, who enrolled in kindergarten just one year after commencing treatment. During her first year at school, she missed no more school days due to colds or other infectious diseases than did her siblings or classmates. In 2009, successful results of another gene therapy trial involving ten children with ADA deficiency were reported: four years after the gene therapy, blood cells in eight of the ten children continued to produce ADA at levels sufficient for good health (Aiuti et al. 2009). Despite these successes, gene therapy for ADA deficiency remains experimental. The preferred way of treating ADA deficiency is a bone marrow transplant from an immunologically compatible sibling or close relative.

All somatic cell engineering trials have not been as successful as those for ADA deficiency. One highly publicized case involved the tragic death of eighteen-year-old Jesse Gelsinger in 1999 during a clinical trial to treat a genetic defect in liver cells. Jesse died a few days after viral vectors carrying copies of a therapeutic gene were infused into his blood.[2] The cause of death was multiple organ failure, apparently due to an allergic reaction to the viral vector. Even though the vector was made noninfectious before the trial, Jesse's immune system responded rapidly and traumatically.

Despite the tragic outcome of Jesse Gelsinger's gene therapy trial and some other equally sad setbacks, clinical trials using somatic cell gene therapy continue. Currently, there are more than fifteen hundred approved, ongoing, or completed clinical trials worldwide for conditions ranging from cancers and cardiovascular diseases to neurological and ocular diseases.[3] In less than a decade, type 1 genetic engineering may offer cures or effective treatments for many devastating diseases and other conditions that are now incurable, including adult-onset diabetes, sickle-cell anemia, familial hypercholesterolemia, certain hereditary immunodeficiencies, Parkinson's disease, some cancers, HIV infection, cystic fibrosis, arterial disease, rheumatoid arthritis, and color blindness.

There is widespread public support in the United States for somatic cell gene therapy. A 1992 Harris poll of one thousand adults showed 88 percent approving of gene therapy to treat or cure genetic diseases. Official statements from international and national councils, commissions, and committees of experts in ethics, biology, and theology also support type 1 genetic engineering.

Germ-line Gene Therapy (Type 2)

Germ-line genetic engineering is routinely performed in agricultural plants and animals. Therapeutic germ-line engineering of humans would mean repairing or replacing disease-causing genes in cells at the very beginning of an individual's life—in eggs or in cells from embryos containing just a few cells (fig. 6.3). Genetically modifying cells at such an early stage of development allows the modification to be passed on to virtually all of the future individual's cells, including eggs or sperm. Subsequent generations would therefore inherit the genetic alteration. Few persons oppose developing germ-line gene therapy to rid future generations of diseases like cystic fibrosis and Tay-Sachs. However, the prospect of applying type 2 genetic engineering to non-disease traits raises several ethical concerns.

How is germ-line genetic engineering done? Two methods are used to genetically alter the germ line in non-human mammals such as mice.[4] Similar methods will likely someday be applied to humans. The first method is to inject copies of the gene(s) of interest directly into the nucleus of egg cells either before or just after fertilization. New genes incorporated into the genome at the egg stage of development end up in all cells of the resulting individual (fig. 6.3). The second method is to genetically engineer *pluripotent* cells taken from the *inner cell mass* of a *blastocyst*-stage embryo or from an *embryonic stem cell (ESC) line.* The engineered cells are then transplanted into the inner cell mass of a blastocyst-stage embryo. The result is an embryo comprised of a mixture of genetically engineered and non-engineered cells. If some of the genetically engineered cells give rise to eggs or sperm, the altered gene(s) can move into future generations.

Under what circumstances might germ-line gene therapy be done in humans? Suppose that a couple wishes to have children biologically related to them but that both partners carry a *recessive genetic disorder* that will be manifested in all of their offspring. This could be the case for persons with cystic fibrosis or sickle-cell anemia. Germ-line therapy could allow a couple like

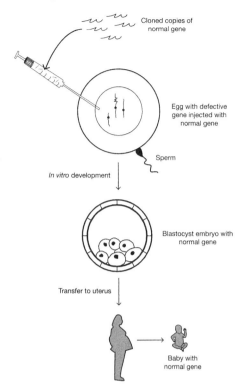

Fig. 6.3. Germ-line gene therapy via genetically engineered eggs. Copies of a normal gene are injected into an egg with a defective copy of the same gene. One or more copies of the normal gene insert themselves into the chromosomal DNA of the egg. The egg is fertilized, allowed to develop in vitro to the blastocyst stage, transferred to a woman's uterus, and carried to term. All of the cells in the baby contain the new gene, including cells that will produce sperm and eggs. The same procedure could be used for germ-line genetic enhancement.

this to have children free from the disorder. Or consider a diagnosable genetic disorder carried by prospective parents that could be life-threatening in some of their children. An example is germ-line retinoblastoma, a devastating genetic disease that leaves victims with multiple types of cancer throughout the body in later life. Germ-line gene therapy could rid all subsequent generations of this family from the dread disease. There are other situations similar to the two just described in which the only feasible way for parents to perpetuate their genetic lines with healthy offspring is via germ-line gene therapy.

Somatic Cell Genetic Enhancement (Type 3)

In May 2001, the *New York Times* reported that athletes will soon seek a competitive edge from altered genes, if they are not already doing so. Injection directly into body tissues of genes for hormones that stimulate production of red blood cells or increase muscle mass and strength could produce performance-enhancing effects lasting for months or even years, compared to the minutes- or hours-long effects achieved by taking the hormone directly. Aside from the fairness issue, genetic enhancements like these pose serious health dangers, including stroke, heart attack, and possibly other biological consequences yet unknown.

In other instances, type 3 genetic interventions may be less objectionable. Currently, unusually short children producing normal levels of human growth hormone (hGH) can enhance their stature by several inches by taking supplemental hGH.[5] Once safety issues are resolved, height enhancement might be accomplished more effectively and economically by simply adding the gene for hGH to some of the children's own cells and reintroducing the engineered cells into the body, where they would produce the hormone and enhance bone and muscle growth. Similar procedures could someday be used to improve memory or cognitive abilities. It is not hard to imagine cases where a particular treatment might be considered therapy for one person (correcting a severe growth defect) but enhancement for another (increasing a child's likely adult stature from five feet to seven feet). Designating a particular genetic intervention as therapeutic or enhancing raises the difficult problem of defining what is normal and what is defective, a problem discussed in chapter 9 and for which there is no clear answer.

Germ-line Genetic Enhancement (Type 4)

Germ-line enhancement is the most controversial type of human genetic engineering. The main concern is that type 4 genetic alterations are not essential for healthy living and they would affect future generations without their consent. Presuming that humankind does eventually undertake germ-line genetic enhancement, what traits might be selected for enhancement?

In *Ethics of Human Gene Therapy*, Walters and Palmer (1997) suggest grouping genetic enhancements into three spheres: physical, intellectual, and moral. Physical enhancements relate to anatomical or physiological characteristics, including longevity and physical well-being; intellectual enhancements relate to traits like memory and intelligence; moral enhancements relate to one's attitudes or behavior toward other persons.

One can imagine physical genetic enhancements that are either health related or purely cosmetic. A health-related enhancement would be engineering the immune system to strengthen its surveillance and responsiveness toward pathogenic viruses and bacteria. Other possible health-related enhancements include decreasing human sensitivity to certain allergens or the harmful effects of ultraviolet light. By contrast, cosmetic enhancements might include preventing age-associated farsightedness, hair loss, or wrinkled skin. Other physical genetic enhancements might someday be incorporated into the human germ line as simple conveniences: decreasing sleep dependence (great for college students and new parents), increasing longevity to "correct" the shorter life expectancy of males, increasing tolerance to heat and cold, or diminishing negative side effects of post-menopausal life.

Intellectual characteristics that could be targeted for enhancement include short- and long-term memory, analytical abilities, attention span, language abilities, and art appreciation. And moral enhancements might focus on temperaments like aggressiveness, kindness, empathy, altruism, and even religiosity. Most intellectual and moral traits are probably controlled by several different genes whose activities are finely coordinated with environmental factors, so engineering these traits would be a complicated undertaking. Of the intellectual and moral traits just named, only aggressiveness shows evidence of being controlled by a single gene.[6] Realistic possibilities for engineering most intellectual and moral traits into the human germ line lie in the indefinite future, but it is not too soon to think about the ethical implications of such an undertaking.

Human Genetic Engineering: Human Values

Primary benefits of human genetic engineering are healthier, happier people. Somatic cell gene therapy shows great promise for treating persons who inherit genetic diseases or acquire gene-based diseases like cancers during their lifetimes. Germ-line gene therapy could prevent genetic disease before birth. Both somatic and germ-line genetic enhancements could make people happier by expanding a person's range of possible life plans or helping one to live his or her conception of a good life.

Despite the potential benefits of human genetic engineering, many objections are raised against the technology. Most of these are against genetic enhancements (as opposed to therapies), especially germ-line enhancements. Examining some of these objections may help to clarify our personal views on the use of powerful genetic technologies to reshape human beings.

Objection 1. Using gene therapy to cure diseases is a proper role for medicine, but enhancing human traits beyond what is normal is "playing God," and that is wrong.

This objection raises at least three different issues: the role of medicine, the distinction between therapy and enhancement, and the sanctity of the human genome. Nearly everyone believes that a proper role of medicine is to treat disease, but beyond that there is little agreement on what boundaries ought to define the practice of medicine. Helping people to obtain and maintain the mental and physical health necessary to live a good life, while doing no harm to others, seems to me a reasonable description of the overarching goal of medicine. But what is needed to live a "good life" can vary between persons. For many a good life may simply require being free of disease and pain, while professional athletes may need extraordinary physical strength or agility, chess players an extraordinary memory, artists extraordinary creativity, and models and screen stars idealized figures and faces. Making a very large nose smaller may be essential therapy for a model and merely cosmetic enhancement for me. There seem to be no clear boundaries that define the proper practice of medicine and apply equally well for every situation.

Opposing genetic engineering for non-therapeutic purposes presumes that one can clearly distinguish between therapy and enhancement. Some contend that therapy corrects what is abnormal and enhancement modifies what is already normal. University of Oxford ethicist Julian Savulescu

(2007, 516) defines enhancement as technology to help people "live a longer and/or better life" than normal, and ethicist Thomas Murray (2007) of the Hastings Center uses the dictionary definition of *enhance:* to advance, augment, elevate, heighten, or increase.[7] Religious studies and Christian ethics scholar Gerald McKenny (2002) acknowledges that human "enhancement" is an imprecise term that can refer to a range of phenomena, including increased capabilities and cosmetic alterations. For Murray, the question is not whether enhancement per se is good or bad, but how others are affected by the enhancement and what values and objectives the enhancement promotes. For example, would a decision by parents to genetically enhance their male embryo to grow into a seven-foot-tall, well-coordinated, athletic man be mainly self-serving or for the well-being of their son?

Shortness due to pituitary gland malfunction is considered abnormal, the pituitary gland in such a person being considered diseased. Another person may simply have shortness in her genes and nothing wrong with her pituitary gland. If neither takes supplemental human growth hormone (hGH) and both grow to be only 4 feet 10 inches tall, is the former person diseased and the latter person healthy but just short? The US Food and Drug Administration indirectly answered this question in 2003 when it approved the use of Humatrope injections for children with idiopathic short stature; that is, with shortness not caused by pituitary gland malfunction or other abnormal conditions, including Turner and Prader-Willi syndromes.[8] With this decision, the FDA effectively defined as abnormal the shortest 1.2 percent of children in the United States, regardless of the causes for their shortness (Drug Information Online 2003). Such "diagnostic creep," where a condition previously considered normal gradually comes to be considered abnormal simply because the technology exists to change it, is also a concern for neuroenhancement via drugs like Prozac and Ritalin (chapter 9). Although the hGH example does not involve human genetic engineering, it highlights the problem of distinguishing between therapy and enhancement.

The "playing God" objection applies to only some notions of God, presumes that one knows the proper role of God, and presumes that human activity should not overlap with God's activities. Most of modern medicine, including surgery, immunizations, assisted reproduction, and pharmaceutical use, is open to the criticism that humans are playing God in the sense that the interventions change the outcome of lives from what they would be without the interventions.

Often the unstated view that the human genome, disease carrying or not, is sacred lies at the root of opposition to human germ-line genetic engineering. That germ-line genetic engineering violates the sanctity of the human genome is raised by some Protestant and ecumenical Christian denominations, but not by Roman Catholic or Jewish traditions. McKenny believes that differing views on human nature and dignity are the reason. According to McKenny, Roman Catholic theology stresses respect for all human life and union of body and soul at procreation, while the Jewish tradition gives humans, and especially physicians, a divine license to heal, repair, and perfect the world. By contrast, conservative Protestant denominations focus on God as the literal creator of all life and describe human nature in terms of God's own image; for them, the human genome is not a dynamic entity with an evolutionary history still unfolding. Rather, they see the genome as a sacred creation fundamental to human life, which reflects God's image. Permanently altering the human genome by germ-line genetic engineering would violate the nature of God itself. Finding that DNA is infinitely alterable and interchangeable between species is problematic for this view of human nature. Whether religious doctrine will be as flexible as the DNA it reveres is yet to be seen.

Objection 2. Although germ-line gene therapy or enhancement per se is not immoral, the research needed to make it safe for humans will inevitably require experimentation on and sacrifice of human embryos, and that is immoral.

Safe germ-line genetic engineering would certainly require research using human eggs and early embryos grown *in vitro* through the *blastocyst* stage and never transferred to a woman's uterus. This objection is significant and intractable for those who assign full personhood to all fertilized eggs and early embryos. It is the same concern that fuels opposition to human

ESC research, discussed in chapter 2. Persons with this view of person-hood may receive consolation from the knowledge that US law prohibits federal (but not private or state) funds from being used to create or destroy human embryos.

Objection 3. Germ-line genetic engineering would violate the autonomy of the unborn individuals subjected to the procedure.

Infringement upon the autonomy of unborn individuals is of greater concern for germ-line enhancement rather than for unambiguous germ-line therapy to prevent disease. The latter would give future individuals more freedom to live a good life. However, germ-line enhancement for specific talents such as musicianship or athleticism could imperil parent-child relationships and jeopardize children's independence to thoughtfully select the talents they wish to develop. However, some germ-line enhancements, such as improved memory, sociability, empathy, or persistence, are consistent with autonomy and could expand children's range of life plans.

In the end, technology may soften this objection by devising ways to inactivate some germ-line enhancements. If artificial chromosomes that can be inactivated were used to genetically alter the human germ line, it is possible that certain enhancements could be removed at the decision of the person carrying them. This would restore genetic autonomy to persons who had received genes for genetic enhancement prior to birth.

Objection 4. Germ-line genetic engineering replaces the mechanism of natural selection with the vagaries of human desire, which are unnatural and therefore unethical.

During nearly all of bipedal primates' several-million-year history on Earth, natural selection has been the most important shaper of our physical and intellectual traits. Only with modern medicine has humankind taken an active role in its own evolutionary history, permitting some genes to remain in the population that probably would otherwise have disappeared. Widespread human germ-line enhancement would allow us to take an unprecedented degree of control over our future evolution. Persons who object to human germ-line genetic enhancement because it is unnatural are, knowingly or not, placing humans and human activity outside the realm of nature. Yet all facets of modern biology tell us that we are a product and a part of nature.[9] Moreover, persons with this objection are hard pressed to explain how germ-line enhancement is any less natural than flying across oceans in metal cylinders or replacing human heart valves with those from pigs.

A variation on this objection is to say that natural selection is wiser than human desire when it comes to designing genomes, so we should use restraint in applying germ-line enhancements to the human genome. For example, augmenting traits like aggressiveness and group loyalty or decreasing the capacity for empathy may have benefits for soldiers or rugby players but may not be in the long-term interest of the species.

Natural selection is sometimes "wise" and sometimes not. Evolution works with preexisting structures rather than engineering innovations from scratch. Therefore, we have undesirable traits such as an appendix prone to infection, wisdom teeth, and a spinal column that fosters lower back problems. Nevertheless, evolution via natural selection is wiser than humans in seeing the big picture. Plant, animal, and microbe components of communities, entire ecosystems, and the global biosphere itself fit together in a complex network of interdependency that works. By contrast, we humans tend to behave with relatively short-term gains in mind, and we often fail to think about how our actions affect the network of interdependent organisms and processes necessary for all living things to thrive. The dangers in humankind taking charge of the structure and evolution of entire ecosystems are discussed in the context of the new discipline of *synthetic biology* in chapter 10. In preparation for controlling our own evolution, which will almost surely come to pass over the next century or two, we now need to nurture in our children and great-grandchildren farsightedness in their desires for themselves and for the species.

Objection 5. We do not know enough about gene function to safely alter the human germ line; in our ignorance, we may cause more harm than good.

This objection is a safety concern rather than a moral objection to the technology per se. The concern is valid because molecular biologists regularly discover new functions for genes previously known to have only one function. So it is possible that in our attempts to alter or delete a gene thought to be only deleterious, we may inadvertently affect other important, unknown functions of the gene. One well-known example of a deleterious gene that also has a beneficial function for some persons in some situations is the gene for sickle-cell anemia. In a single dose, the gene helps protect individuals against malaria.

On the other hand, a modified gene that causes fatal or debilitating disease in a single dose, that is, a dominant genetic condition, such as that causing Huntington disease, can have no hidden adaptive advantage. Ridding our genomes of these kinds of genes and replacing them with normal copies of the disease-causing genes seems risk free. The prudent course is to be humble about our knowledge of gene function and our adeptness at improving what we have inherited via natural selection. When germ-line genetic engineering is finally applied to humans, it will probably be restricted for many decades to removing or repairing only those genes known to cause fatal or debilitating conditions.

Objection 6. Using genetic enhancement technologies will promote a dangerous, new eugenics that stigmatizes and devalues those with certain genetic traits or those who lack enhancements.

In his 1883 book, *Inquiries into Human Faculty,* Francis Galton drew on the theory of natural selection proposed by his half-cousin, Charles Darwin, and suggested that the human species can be improved through better breeding. At first, applications of positive eugenics in Britain and the United States encouraged persons with attractive features and strong physiques to marry and produce children. However, as we saw earlier (chapter 3), within a few decades negative eugenics programs emerged in the United States, Canada, and Germany with dire consequences for human rights and human life. With the advent of human genetic biotechnologies,

Table 6.2. Types of eugenics

	Positive	Negative
Individual	I	II
Population	III	IV

Note: Two ways of classifying eugenics, positive versus negative and individual versus population, give rise to four types of eugenics.

conversations about eugenics have resumed. To engage in these conversations, we must agree on some terminology.

University of Pennsylvania bioethicist Arthur Caplan (2000) identifies two ways to categorize eugenics. The first is the positive-negative dichotomy. Positive eugenics actively promotes the reproduction of individuals with desirable genetic-based traits, while negative eugenics aims to eliminate deleterious or undesirable genes from the gene pool. The second categorization distinguishes between "individual" and "population" eugenics. Individual eugenics is practiced by individual couples without the aim of having a specific effect on the societal gene pool. By contrast, population eugenics programs are institutionalized efforts to improve the overall gene pool of large populations. These two ways to categorize eugenics generate a matrix analogous to the one used earlier to classify types of genetic engineering, resulting in four types of eugenics (table 6.2).

Caplan (2000) favors the freedom of parents to make choices about the genetic endowment of their offspring (eugenics types 1 and 2). He supports individual eugenics so long as the methods pose no health risks for children and parental decisions do not compromise children's life opportunities. Caplan believes that parents' decisions about the genetic traits of their children are fundamentally no different from decisions to expose children to positive environmental influences such as libraries, piano lessons, and organized sports. On the other hand, Caplan opposes population eugenics. Attempts to improve the gene pool of populations, he fears, could too easily shift toward imposition of institutional visions of perfection upon individuals, with dire consequences. His fear is well founded, as twentieth-century genocide, racism, and wars of ethnic cleansing are all traceable to efforts at implementing such institutional visions of perfection. Still, many of us might support government programs using safe germ-line genetic engineering to make everybody's offspring resistant to HIV infection, Alzheimer's disease, or cardiovascular disease. The problem is in the fuzzy area between disease therapy or prevention and enhancement. Where should we draw the line between these two types of genetic engineering, especially when public funds are involved?

Not surprisingly, a wide range of thoughtful, informed opinions exist on whether humankind ought to take control of its own evolution using germ-line genetic engineering. Francis Fukuyama, former member of George W. Bush's President's Council on Bioethics, opposes germ-line genetic enhancements, fearing they will further stratify society. He argues that the technology may lead to the emergence of a powerful ruling class so advanced over the unenhanced that constitutional democracy would break down. By contrast, molecular biologist Gregory Stock sees human germ-line engineering as inevitable and believes we ought to concentrate on making wise decisions as we proceed ahead with it rather than miring ourselves in debates over whether to do it at all.

Objections 7 and 8. Unequal access to technologies of genetic enhancement will further stratify society into the "haves" and "have nots," and human genetic enhancement research will divert efforts and funds from projects to help persons worldwide who now lack basic needs for a good life.

McKenny, an applied ethicist, identifies these as two of the most important ethical issues facing human genetic engineering, and he recommends setting aside the "tired debate over whether religious traditions should support gene therapy" (2002, 289).[10] Distributive justice and biomedical research priorities, important issues for all twenty-first-century biotechnologies, are discussed at the end of chapter 2.

Conclusion

Two means for classifying human genetic engineering are therapeutic versus enhancing and somatic (transient) versus germ line (permanent). No type of genetic engineering is yet a routine medical practice, but by 2010 at least fifteen hundred clinical trials of somatic cell gene therapy were completed or underway. The concept of somatic cell gene therapy is relatively uncontroversial, although some tragic failures of the technology raise safety issues. Genetic enhancement and human germ-line genetic engineering, on the other hand, raise several ethical concerns. Religion-based objections to germ-line and enhancing engineering include the belief that humans should not "play God" and that the human genome is sacred and should not be tampered with. Secular-based objections to germ-line and enhancing engineering include violation of the autonomy of the unborn, their possible eugenic uses, and potential negative effects on society due to disparities in their accessibility. Some ethicists believe we should refrain from using genetic technologies for enhancements altogether, whereas others believe their use is inevitable. One confounding element in these discussions is difficulty in clearly distinguishing between what is therapy and what is enhancement.

Questions for Thought and Discussion

1. Do you believe gene therapy and genetic enhancement are distinguishable from each other? If so, by what criteria?
2. What do you believe are the five most important qualities a person needs to live a good life? Should genetic enhancements be applied to giving these qualities to all people?
3. Is there a qualitative difference between parents nurturing their children with attention, nutrition, and a loving environment and nurturing their children with genetic enhancement aimed at improving their lives? In thinking about this question, assume that attention, nutrition, and love all have lasting effects on children's epigenomes.

4. Assuming it can be done safely, do you favor using germ-line gene therapy to prevent diseases like sickle-cell anemia, cancers, Alzheimer's disease, and diabetes in present and future generations?

5. Assuming it can be done safely, do you favor using genetic engineering to reduce propensities for aggression and criminal behavior, social anxiety, alcoholism, neuroticism, and parental neglect?

6. Should the appropriateness of genetic enhancements in sports and professional life be judged differently from currently allowed performance enhancers such as caffeine and protein supplements?

7. Assuming it can be done safely, would you favor using genetic engineering to increase human happiness? If you oppose genetic engineering for happiness under any conditions, why do you oppose it? What is happiness?

8. Do you believe that genetic engineering will be a major factor in the future evolution of *Homo sapiens?* If so, what directions do you see human evolution taking? If not, why not?

Notes

1. ADA is adenosine deaminase. The gene for ADA is on chromosome 20, and ADA deficiency causes an accumulation of substances that are toxic to immature cells in the developing cellular immune system; therefore, the immune system in ADA-deficient patients is very weak or nonexistent.

2. The therapy was intended to correct a deficiency for the enzyme ornithine transcarbamylase. This enzyme deficiency causes a toxic accumulation of ammonia in the blood that can lead to mental retardation, abnormal growth, seizures, and coma.

3. See the "Gene Therapy Clinical Trials Worldwide," provided by the *Journal of Gene Medicine,* at http://www.wiley.co.uk/genetherapy/clinical/ (accessed October 2, 2009). This site gives numbers of clinical trials by disease category, gene transferred and vector used, trial country, clinical phase, trial status, and year approved and initiated from 1989 to the present.

4. Germ-line genetic engineering of mice is used extensively to create mice lacking particular genes in order to learn indirectly about the function of those

genes; these are called gene "knock-out" mice. Gene "knock-in" mice are also created by adding one or more copies of particular genes to gain insight about their functions.

5. Many factors besides a normal level of growth hormone contribute to stature. In some cases above normal levels of hGH will compensate for some of those other factors and enhance stature.

6. This evidence comes from genetic studies on a Dutch family, indicating that a mutation in the gene for an enzyme that regulates levels of certain brain chemicals may be the biological cause of abnormally violent behavior.

7. The Hastings Center, founded in 1969, is a nonpartisan research institution dedicated to bioethics and the public interest.

8. Humatrope is the commercial name for human growth hormone, or somatotropin, manufactured by Ely Lilly Company using genetically engineered bacteria containing the human gene for growth hormone.

9. Neil Shubin's (2009) book, *Your Inner Fish: A Journey into the 3.5-Billion Year History of the Human Body,* is a very accessible and entertaining explanation of how human morphology and genetics reveal our species' intimate, ancestral connections with other living things.

10. I find this recommendation especially significant since it comes from a Christian ethicist very familiar with the religion-based objections to human genetic engineering.

Sources for Additional Information

Brunner, H. G., M. Nelen, X. O. Breakefield, et al. 1993. "Abnormal Behavior Associated with a Point Mutation in the Structural Gene for Monoamine Oxidase A." *Science* 262: 578–80.

Caplan, A. L. 2000. "What's Morally Wrong with Eugenics?" In *Controlling Our Destinies,* edited by P. Sloan, 209–22. Notre Dame, IN: University of Notre Dame Press.

Chapman, A. R., and M. S. Frankel. 2003. "Framing the Issues." In *Designing Our Descendents: The Promises and Perils of Genetic Modifications,* edited by A. R. Chapman and M. S. Frankel, 3–19. Baltimore, MD: Johns Hopkins University Press.

Fox, D. 2005. "Human Growth Hormone and the Measure of Man." *New Atlantis* (Fall 2004/Winter 2005): 75–87.

Fukuyama, F. 2002. *Our Posthuman Future: Consequences of the Biotechnology Revolution.* New York: Farrar, Staus & Gieroux.

Garreau, J. 2005. *Radical Evolution: The Promise and Peril of Enhancing Our Minds, Our Bodies—and What It Means to Be Human.* New York: Doubleday.

Gilbert, S. F., A. L. Tyler, and E. J. Sackin. 2005. *Bioethics and the New Embryology.* Sunderland, MA: Sinauer Associates.

Goldwert, L. 2007. "Hope for Sufferers Of Parkinson's Disease: Experimental Gene Therapy Shows Promise for Easing the Symptoms of Degenerative Disorder." CBS News, June 21. http://www.cbsnews.com/stories/2007/06/21/health/main2964032.shtml (accessed October 1, 2009).

Highfield, R. 2007. "Breakthrough in Parkinson's Gene Therapy." Telegraph.co.uk, November 19. http://www.telegraph.co.uk/science/science-news/3314942/Breakthrough-in-Parkinsons-gene-therapy.html (accessed October 1, 2009).

Jonsen, A. R. 1998. *The Birth of Bioethics.* New York: Oxford University Press.

Journal of Gene Medicine. "Gene Therapy Clinical Trials Worldwide." John Wiley & Sons Ltd. http://www.wiley.co.uk/genetherapy/clinical/ (accessed October 1, 2009). This site gives numbers of clinical gene therapy trials by year from 1989 to the present, continent, country, type of gene transferred, condition treated, vector used for the gene transfer, and clinical phases I–III.

Kaplitt, M. G. 2007. "Safety and Tolerability of Gene Therapy with an Adeno-Associated Virus (AAV) Borne GAD Gene for Parkinson's Disease: An Open Label, Phase 1 Trial." *Lancet* 369: 2097–2105.

McKenny, G. P. 2002. "Religion and Gene Therapy: The End of One Debate, the Beginning of Another." In *A Companion to Genethics,* edited by J. Burley and J. Harris, 287–301. Molden, MA: Blackwell Publishers.

Murray, T. H. 2007. "Enhancement." In *The Oxford Handbook of Bioethics,* edited by B. Steinbock, 491–515. New York: Oxford University Press.

Sandel, M. J. 2007. *The Case against Perfection: Ethics in the Age of Genetic Engineering.* Cambridge, MA: Belnap Press of Harvard University Press.

Savulescu, J. 2007. "Genetic Interventions and the Ethics of Enhancement of Human Beings." In *The Oxford Handbook of Bioethics,* edited by B. Steinbock, 516–35. New York: Oxford University Press.

Shoubridge, E. A. 2009. "Asexual Healing." *Nature* 461: 354–55. This article summarizes the results of the technical study on mitochondrial gene replacement reported by Tachibana and coworkers.

Stock, G. 2002. *Redesigning Humans: Our Inevitable Genetic Future.* New York: Houghton Mifflin Co.

Tachibana, M., M. Sparman, H. Sritanaudomchai, et al. 2009. "Mitochondrial Gene Replacement in Primate Offspring and Embryonic Stem Cells." *Nature* 461: 367–72.

Wade, N. 2009. "With Genetic Gift, 2 Monkeys Are Viewing a More Colorful World." *New York Times,* September 22. http://www.nytimes.com/2009/09/22/science/22gene.html?pagewaned=print (accessed October 2, 2009).

Walters, L., and J. G. Palmer. 1997. *Ethics of Human Gene Therapy.* New York: Oxford University Press.

7
Human Reproductive Cloning
Sameness, Uniqueness, and Personal Identity

The cloning of humans is on most of the lists of things to worry about from Science, along with behaviour control, genetic engineering, transplanted heads, computer poetry and the unrestrained growth of plastic flowers.

—Lewis Thomas

This chapter is about the prospect of generating genetic copies of preexisting human beings, *reproductive cloning.* It differs from *therapeutic cloning (embryo cloning),* discussed in chapter 2, in that cloned embryos would actually be transferred to a woman's uterus and brought to term.

Most people fear and oppose human cloning. Gallup poll results, reported by the California-based Center for Genetics and Society, show that 85–96 percent of US and European Union citizens and 71–93 percent of Eastern Europeans oppose human reproductive cloning.[1] Before the 1978 birth of Louise Brown, the first test-tube baby, many people predicted that in vitro fertilization (IVF) babies would be deformed, monster-like, or otherwise terribly abnormal. Now thousands of normal, healthy babies are born every year after IVF. Might the same someday be said about human clones?

Despite several hoaxes and false claims over the past thirty years or so that human clones had been made, as of this writing (early 2012), there is no evidence that any diploid (with the normal number of chromosomes) human has ever been cloned, either to the embryo stage or to full term. But cloning technology in mammals has advanced steadily since 1997, when Scottish cell biologists announced the birth of a cloned lamb, the first mammal ever to be cloned. In 2007, cloned monkey embryos were created (Byrne et al. 2007), and it is likely that similar procedures will work in humans.[2] So public understanding about cloning technology and dialogue about the wisdom of moving ahead with human cloning are urgently needed. Toward these ends, we consider the following specific questions:

1. What does basic research in *cell* biology have to do with cloning technology?
2. How do plants and animals normally reproduce, and how does cloning differ from normal sexual reproduction?
3. How is cloning done?
4. Does human cloning pose biological risks?
5. What moral arguments are there against cloning, and how valid are they?
6. How do the world's religions view human cloning?
7. What laws regulate human cloning?

Cloning: The Biology

In this section we consider the decades-long history of animal cloning, how cloning is related to nature's ways of reproducing organisms, and some of the biological risks of human cloning.

Animal Cloning: Historical Background

In the early 1960s, biologists discovered that virtually every *somatic cell* of an organism has the same genetic information. They called this phenomenon DNA equivalency. DNA equivalency begins with the two cells produced when a fertilized egg (the *zygote*) undergoes its first cell division, and it continues between the trillions of cells an animal generates in its lifetime. The meaning of DNA equivalency is that nearly every cell in the body contains the same information that was originally present in the fertilized egg. With knowledge of DNA equivalency, biologists recognized the theoretical possibility of growing entire organisms using the DNA inside virtually any body cell.

Despite the theoretical possibility of cloning, most biologists were initially skeptical about actually performing such a feat. The cause for this skepticism was that for a specialized somatic cell to perform its specialized functions, most of its genes must be kept inactive. For example, the genes for muscle fiber proteins are active inside muscle cells, but genes for other proteins, such as hair, insulin, hemoglobin, growth hormone, and gut digestive enzymes, are turned off in muscle cells. Restoring complete genetic potential (*totipotency*) to somatic cell DNA and reactivating thousands of genes in just the right sequence to orchestrate the development of a new organism seemed beyond the reach of laboratory science. But science and living cells do amazing things, and the doubters were wrong.

In 1966 two developmental biologists, John Gurdon at Oxford University and V. Uehlinger at the University of Geneva, reawakened the *developmental potential* in the DNA of a single cell from a tadpole to produce an entirely new tadpole. They did this by transferring the DNA-containing nucleus from a cell in the tadpole's gut into an egg cell denuded of its own nucleus. When the gut cell nucleus found itself surrounded by the environment of the egg cytoplasm, its DNA began to behave like the DNA in a newly fertilized egg. The DNA replicated and the host egg cell divided into two cells. More cycles of cell division followed, producing a frog embryo that eventually grew into a normal tadpole, a genetic clone of the tadpole that donated the somatic cell DNA.

The work of Gurdon and Uehlinger (1966) demonstrated that totipotency is inducible in highly specialized cell types. It also opened new avenues of research to learn how cells become functionally specialized during normal development, knowledge with relevance for preventing and treating cancer and preventing birth defects. Reports of cloned mice and even a cloned person in the 1970s turned out to be bogus. But on February 23, 1997, Scottish reproductive biologist Ian Wilmut jolted biology and humanity to the reality of mammalian cloning (Wilmut et al. 1997). On that day he introduced the world to Dolly, a seven-month-old lamb created from the genetic information in a single cell from an adult ewe (fig. 7.1).

As with the cloned frog, Dolly was created by introducing the nucleus from a somatic cell into an *enucleated egg,* one from which the nucleus has

Fig. 7.1. Dolly, the cloned sheep, and her lamb, Bonnie. Copyright by the Roslin Institute, University of Edinburgh, Scotland, printed with permission.

been removed (fig. 7.2) This method of cloning is called *somatic cell nuclear transfer (SCNT)* (fig. 2.8). Since Dolly's birth, SCNT has produced cloned mice, cows, goats, pigs, a bull, an endangered Asian ox, a cat, a race mule named Idaho Gem, dogs, and monkeys.

As the first mammal ever cloned, Dolly became an overnight celebrity. Genetic tests confirmed the claim of Dolly's creators, showing that she possessed DNA identical to DNA from the donor ewe. Humans are mammals and share many aspects of their reproductive biology with sheep, mice, and other hairy creatures. Therefore, with some refinements, the technology used to create Dolly can probably be used to clone humans.

In the days and weeks after the Dolly announcement, the media catapulted the specter of cloned people from the realm of science fiction into reality. In an interview shortly after the announcement of Dolly's birth, Dr. Wilmut was asked about the uses of cloning. Wilmut spoke of the value of cloned animals for medical research and dismissed the idea of cloning humans, saying, "All of us would find that offensive" (quoted in Kolata 1997). Wilmut has also suggested that an international moratorium on

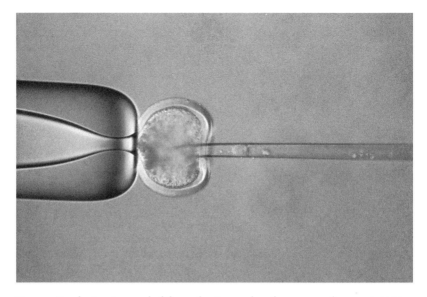

Fig. 7.2. Enucleation (removal of the nucleus) procedure for a mammalian egg (*middle*) using a hollow glass needle (*right*). A suction device (*left*) holds the egg in place as the researcher pierces the egg with the needle and sucks out the nucleus, which is visible inside the needle just outside the egg. Photograph used with permission from the Roslin Institute, University of Edinburgh, Scotland.

human cloning should be put in place. As politically correct as Wilmut's words are, the public is still anxious over the prospect of human cloning. Now, more than fifteen years post-Dolly, no international moratorium on human reproductive cloning yet exists, and no federal law completely prohibits human cloning in the United States. But valuable discourse on the issue has occurred, beginning with a letter from President Bill Clinton to Dr. Harold Shapiro, chair of the National Bioethics Advisory Commission (NBAC), on the very day of Dolly's birth announcement:

Dear Dr. Shapiro:

As you know, it was reported today that researchers have developed techniques to clone sheep. This represents a remarkable scientific discovery, but one that raises important questions. While this technological advance could offer potential benefits in such areas as medical research and agriculture, it also raised serious ethical questions, particularly with respect to the possible use of this technology to clone human embryos. Therefore, I request that the National Bioethics Advisory Commission undertake a thorough review of the legal and ethical issues associated with the use of this technology, and report back to me within ninety days with recommendations on possible federal actions to prevent its abuse.

Sincerely,

Bill Clinton[3]

Dr. Shapiro responded to President Clinton 105 days later with a 125-page report prepared by his eighteen-member commission. The commission consulted ethicists, theologians, scientists, physicians, and concerned citizens to produce a thorough report addressing human reproductive cloning. In this chapter we examine the findings of President Clinton's commission and also other statements and positions on human cloning.

Sexual Reproduction, Asexual Reproduction, and Cloning

How does reproductive cloning differ from the sexual method used by plants and animals for at least one billion years? During *sexual reproduction,* genes from two individuals combine to generate a new individual similar to the parents but not identical to either one. In contrast, during *asexual reproduction,* a single individual produces a new individual genetically identical to itself, without involvement of another's genes. Sexual organisms begin with the union of an egg and sperm. Asexual reproduction happens when an amoeba divides to produce two amoebae, or when a small shoot cut from a plant sprouts roots and generates a new plant. Cloning by SCNT is like the dividing amoeba and the sprouting shoot. A small bit of the body of the original organism gives rise to a new organism genetically identical to the first. In the case of SCNT, the body bit is the DNA-containing nucleus from a single, donor, somatic cell.

Transferring the donor nucleus into an enucleated egg reprograms the donor DNA back to a state of total developmental potential like that in a fertilized egg. Chemical signals in the egg cytoplasm do the reprogramming, allowing the transplanted DNA to rerun the sequence of gene activity needed to form a new individual from a single cell. In essence, SCNT tricks the egg into behaving as though it has just been fertilized.

There are two ways to transfer a somatic cell nucleus into an enucleated egg for cloning: (1) by physically removing the nucleus from a somatic cell and injecting it into the enucleated egg (figs. 2.8 and 7.2) or (2) by simply fusing the entire somatic cell with the egg (fig. 7.3). After transfer of the somatic cell nucleus to the egg cytoplasm, treating the egg chemically or by electric shock activates it to begin dividing. In mammals, a small fraction of the artificially activated eggs behave like newly fertilized eggs and produce normal blastocyst embryos that can be transferred to the uterus for the completion of development.

Biological Risks of Human Cloning

Physical risks to the clone. There is still much to learn about developmental aberrations that might occur due to the cloning process itself. In farm animals and in rhesus monkeys, the survival rates of embryos generated by cloning are less than 3 percent, and most pregnancies fail just before term. In the work that produced Dolly, only 1 of 273 embryos implanted into uteruses of adult sheep resulted in a live birth. Even when cloned mammals are born alive, they often become inexplicably obese, and cloned mice that appear to be normal sometimes have abnormal patterns of *gene expression* in their cells that can cause physiological problems.

Another unknown is how the age of the donor of genetic material affects the life span and health of the clone. Normal mammalian cells divide a limited number of times before they die. Some evidence suggests that regions of DNA at the ends of chromosomes, called *telomeres,* may be part of the counting mechanism. Telomeres shorten a little bit with each round of DNA replication preceding cell division. Eventually, shortening at the ends of chromosomes damages genes vital to the life of the cell. So, cloning

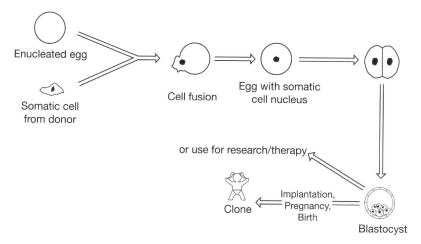

Fig. 7.3. Cloning by somatic cell nuclear transfer (SCNT). Fusion of a diploid somatic cell with an enucleated egg cell produces an egg that mimics a fertilized egg and can develop into a complete individual.

by SCNT using donor cells from adult tissue may shortchange the clone's health and life span.

This could be what happened to Dolly. Analysis of her chromosomes soon after birth revealed abnormally short telomeres. They were about the length of the telomeres in the six-year-old ewe that donated the nucleus used to create Dolly. Before being euthanized in 2003, Dolly suffered from lung problems, arthritis, and other maladies rare in a sheep just six years old. Whether Dolly's health problems resulted from cloning per se or from beginning life with shortened telomeres is not known.

Based on results in animals, if humans were cloned now and brought to term, the clones would probably experience an abnormally high incidence of birth defects and a high risk of developing genetic-based abnormalities later in life. Although the physical risks associated with cloning in mammals may someday be overcome, currently they are reason enough to refrain from the reproductive cloning of humans.

Would cloning be bad for the species? Fears that human cloning would dangerously diminish the gene pool of *Homo sapiens* and thereby make the species less adaptable are not founded in fact. Other species of animals are considered endangered when their worldwide populations fall below one thousand. With billions of sexually reproducing humans in the world, repetition of the genome of even millions of these by cloning would not endanger the gene pool of *Homo sapiens.*

Reproductive Cloning: Human Values

The strongest arguments for human cloning are based on the view that the liberty to reproduce, even if assisted reproduction is needed, is a right protected by the US Constitution (Robertson 1994). Arguments against human cloning are based on values either rooted in secular society or embedded in religious traditions. The greatest secular-based concerns are over the physical and psychological well-being of the clone and the potential undermining of shared social values. Religious concerns are diverse, reflecting the variety of religious traditions in our world society. By examining some of the objections to human reproductive cloning, we will be better able to form and defend our own informed opinions on the subject. You may or may not agree with my analyses, so I urge that you also give your own informed critique of each objection.

Objection 1. Cloning humans is playing God, and it is wrong for humans to play God.

We discussed the "playing God" objection to biotechnology in chapter 6, making the points that all views of God do not preclude using biotechnology to manipulate human biology and that modern medicine itself is also vulnerable to this objection. Rationality requires rejection of this objection as a valid reason to prohibit human reproductive cloning. But maybe there are other reasons to prohibit cloning; let's see.

Objection 2. To make reproductive cloning safe would require preliminary experimentation with human eggs, early embryos, and embryos transferred to women's wombs. Sacrificing blastocysts and endangering women and developing fetuses in the interest of developing a technology for safe reproductive cloning would be morally unacceptable.

Research with early human embryos would certainly be needed in order to make SCNT safe for human reproductive cloning. For persons who assign full personhood to fertilized eggs and embryos through the blastocyst stage, such research would be morally unacceptable. But even if the technology for generating human embryos by SCNT were perfected, trial transfers of cloned embryos into women's wombs would need to be done. A major ethical concern would be how to respect the autonomy of the unborn clone whose health is put at risk. The concern might be addressed as it was for early IVF procedures, by giving parents the authority to make informed choices on behalf of their unborn babies. Nevertheless, this objection is substantive and deserves thoughtful public discussion.

Objection 3. Cloning humans is just plain disgusting, and we should listen to our intuition because it is telling us that reproductive human cloning violates some basic moral principle.

Gut-level disgust against cloning humans is probably the most widespread argument against cloning. Ethicists Leon Kass (1997) and Leigh Turner (2004) weigh in on opposite sides of this objection to cloning. Kass, a member of President George W. Bush's Council on Bioethics, wrote a lengthy article titled "The Wisdom of Repugnance" (1997). Kass argues that gut-level feelings against procedures like human cloning ought to be heeded because they usually reflect inherent wisdom about what is immoral. Turner, on the other hand, refutes Kass's argument in a two-page piece titled "Is Repugnance Wise? Visceral Responses to Biotechnology" (2004). Turner points out that feelings of repugnance are usually learned responses that reflect prejudices and fears rather than wisdom about natural moral laws.

There is wisdom on both sides of this argument. The nearly universal repugnance toward torturing helpless animals or persons, harming children, and betraying trust appears to reflect some inherent wisdom about morality. On the other hand, validating some persons' feelings of outrage would mean rescinding civil liberties for some persons, outlawing certain political parties, and censoring literature, music, and other art forms. Such actions would, of course, incite equally strong feelings of outrage from the other direction, and chaos would result from attempts to respect everybody's sense of disgust.

How then are we to decide when feelings of disgust reflect the violation of a moral principle? In his book *Clones, Genes, and Immortality* (1998), bioethicist John Harris suggests that a moral principle is recognizable by showing that its violation harms or otherwise adversely affects persons, causes injustice, or violates human rights. Applying this criterion to a moral prohibition of human cloning requires one to show how cloning would harm somebody, violate their rights, or cause injustice. Discussing objections 4, 5, and 7 will help us determine whether human reproductive cloning would cause any of these negative consequences.

Objection 4. Every person has the right to a unique genetic constitution, and this right would be violated by cloning.

Persons with this objection may be forgetting that identical twins share identical genetic constitutions. Few would suggest that we should prevent the birth of identical twins in order for everybody to have a unique genetic constitution. In fact, identical twins share more genes than would a clone and the cloned person.[4] So, for both biological and logical reasons, the claim that we all have a right to a unique genetic constitution must be rejected.

Objection 5. Every person has the right to an open future, the freedom to make what one desires out of one's life. Clones may have this right jeopardized due to others' expectations that they will lead a life similar to that of the cloned person.

Nobody knows for certain the relative roles of environmental and genetic factors in determining life's directions, but some psychologists weight it roughly 60 percent environmental and 40 percent genetic. Our genetic constitution provides a substrate upon which our experiences and surroundings act to erect the collage of traits and talents that make each of us unique. A vast array of prenatal and postnatal influences including the diet, health, stress of the mother during pregnancy, and the child's home environment, everyday life experiences, friends, books, television programs, accidents, joys, and sorrows all contribute to forming a unique person. Cloning Pablo Picasso would not produce another Picasso, or even necessarily an artist.

The open future of a clone may be at risk, but not due to the cloning procedure itself. Rather, a clone's open future may be jeopardized by misunderstandings about the role genetics plays in forming persons. Unrealistic expectations from prospective parents of clones, from the clones themselves, and from others are what imperil the psychological freedom of possible future clones. The most logical solution to this problem is improved public education about the roles of environment and genetics in forming unique persons, not an absolute prohibition on cloning.

Objection 6. Cloning removes reproduction from the realm of human sexuality, and this is morally wrong.

This objection is often voiced by persons with a religious belief that sexuality is a divinely bestowed gift for the expressed purpose of procreation. Although this belief usually precludes using most forms of contraception, it does not logically argue against procreation by means other than sexual intercourse. In any case, artificial insemination and IVF are vulnerable to the same objection. Banning cloning because it bypasses sexual intercourse would be inconsistent with allowing most forms of assisted reproduction. A somewhat different formulation of the objection might claim that everybody has the right to two biological parents. But even if society were to agree on such a right, cloning would not violate it because the genetic material used to produce a clone *is* the product of two parents—the biological mother and father of the cloned person.

Objection 7. Cloning could cause a gradual, unplanned, stratification of society, even under non-malevolent governments, into a group of financially, intellectually, or otherwise elite, cloned persons who might maintain political and social control over the less privileged.

This objection raises the concern that human cloning might lead to the purposeful or unplanned emergence of either persons lacking the autonomy and rights possessed by other persons, or persons that threaten the autonomy of the rest of us. The two scenarios are often depicted in films and novels about clones in which societies use clones as slaves or are somehow endangered by clones.

Similar scenarios in real life have unfolded many times during human history without cloning. The existence of underprivileged or exploited groups is unjust, and strong means should be used to rectify it where it still occurs. But cloning does not pose a unique danger in this regard. Cloning as a means for producing large numbers of like people is not feasible because of the vast array of non-genetic influences that contribute to shaping adult individuals. In fact, a malevolent dictator wishing to control masses of people would be better off using a chemical akin to Aldous Huxley's (1932) soma than undertaking mass cloning.

Objection 8. Human cloning violates human dignity.

That human cloning violates human dignity is the broadest and most ambiguous objection to cloning that is commonly cited in policy statements. Sometimes human dignity is equated with having genetic uniqueness (objection 4), enjoying an open future for one's life (objection 5), or entering the world via a natural process (objection 6). At other times human dignity is invoked at the community level, and concerns are raised about the denigrating effects human cloning might have on society (objection 7). In the end, a clear and concise definition of human dignity is elusive, and it is unclear how the act of cloning per se would violate any of the common usages of human dignity.

Is Human Cloning Morally Wrong?

Of the eight commonly voiced objections to human reproductive cloning just examined, none present strong arguments against human cloning per se. The strongest objections, in my view, are that the research required to develop the technology may pose serious health risks for the technology's pioneer fetuses and that unrealistic expectations of clones might compromise their freedom to develop life plans.

Are there other reasons to consider human reproductive cloning unethical? The National Advisory Board on Ethics in Reproduction (NABER) deemed human cloning to be unethical in the following hypothetical instances:[5]

1. When the purpose is to produce embryos for storage in order to create a twin sibling of a child born years earlier.
2. When the purpose is to create embryos for possible use later to grow a fetus in order to provide compatible bone marrow or organ transplants for a sick child.
3. If the ultimate objective is to provide an adult with an identical twin to parent as his or her own child.
4. If the purpose is to use an embryo clone to replace a child that has died.
5. If the purpose is to be able to produce embryos for later sale to others or to produce fetal tissue or organs for donation to others.

Potential Benefits of Human Cloning

Potential biological benefits of human cloning are usually unknown or overlooked in cloning debates. Even if knowledge about the benefits of human cloning does not temper the widespread negative feelings about cloning, information about possible benefits of cloning deserves to be included in the discussion. Here we consider possible benefits of cloning to individuals and society.

Cloning could reduce health risks associated with IVF. Human cloning by embryo splitting could greatly reduce health risks of in vitro fertilization (IVF) for some women.[6] To have a baby by IVF, women take hormones or drugs to induce multiple, simultaneous ovulations. The aim is to obtain a dozen or so mature eggs, fertilize them in vitro, and then transfer two or three healthy embryos to the womb. But IVF carries the risk of ovarian hyperstimulation syndrome, a dangerous condition caused by enlarged ovaries and fluid accumulation due to the hormone treatments. Egg harvesting itself is an invasive procedure and also poses health risks. A small percentage of women undergoing hormone treatments still ovulate only a single egg. If pregnancy is not achieved with the embryo from the single egg, the treatments are repeated, and sometimes even a third round of ovarian stimulation and egg harvesting is needed. Serious health risks occur with each repetition of the egg procurement procedure. For women who obtain only a single embryo after IVF, cloning by embryo splitting could provide six to eight embryos for uterine transfer, plenty for at least three attempts at pregnancy. Cloning would eliminate the need for multiple hormone treatments and surgeries. The chances for single ovulators to achieve pregnancy would be greatly increased while minimizing health risks and expense.

Therapeutic cloning. Biologically, a clone could be of great medical benefit to the cloned individual as a source of lifesaving, immuno-compatible umbilical cord blood or bone marrow cells, whose donation would not harm the clone. Of course, creating a clone solely to benefit another person, such as to be a sacrificial organ donor, would be indisputably immoral. NABER rightly considers it unethical to produce and store cloned embryos as a type of health insurance. But the situation is different when a child with a life-threatening disease, potentially curable with umbilical cord cells from a clone, already exists. Then the circumstances are similar to the savior sibling situations in the Nash and Ayala families (chapter 3). Recall that the savior siblings in these families are loved and cherished for themselves and not just valued for having saved a sibling's life. Nevertheless, a 2001 *Time/ CNN* poll showed that only 21 percent of Americans think it is justifiable to create a human clone to save the life of another person, even if there were little or no danger to the clone itself.

Religious Belief and Human Cloning

What positions on cloning emerge from various religious traditions? Few religious bodies have published positions on human cloning, but one can surmise how a majority of adherents to a particular religious tradition is likely to view human cloning by looking at how the tradition describes the relationship between humankind, nature, and God. Most cultures inherit ancient, cosmogonic myths that explain the origins of the cosmos and human beings and that define the proper relationship between Creator and the creation. The influence of cosmogonic myths in informing a culture's theological response to modern-day moral dilemmas cannot be overestimated.

Courtney S. Campbell (1998), professor of philosophy at Oregon State University, was commissioned by the US National Bioethics Advisory Commission (NBAC) in 1997 to prepare a statement comparing religious views on human cloning. In selections of his findings and in others' views on human cloning that are summarized or quoted below, the influence of cosmogonic myths is apparent.

Judaism and Christianity. Convictions about the meaning and sanctity of human life within Judeo-Christian traditions are especially interesting to consider because a common cosmogonic myth (i.e., biblical accounts of creation in Genesis) gives rise to a variety of viewpoints. These are described in chapter 3 ("Religious Perspectives") of the cloning report of the National Bioethics Advisory Commission (1997, 39–62) requested by President Clinton. Some of that information is incorporated into what follows here.

Roman Catholicism clearly and vehemently opposes human cloning. The Catholic moral tradition views human cloning as a violation of human dignity under any circumstance, and this position is based on the Vatican's interpretation of the Creation story and God's redemptive acts on behalf of individuals. In April 1997, a statement from the Vatican implored all nations to ban human cloning and declared that all persons have "the right to be born in a human way." By "human way" is meant biblically ordained sexual union within the sacred covenant of marriage: "Male and female created he them. And God blessed them, and God said unto them, Be fruitful, and multiply, and replenish the earth" (Genesis 1:27–28). "Therefore shall a man leave his father and his mother, and shall cleave unto his wife: and they shall be one flesh" (Genesis 2:24).

Protestant theologian Joseph Fletcher sees things differently. Fletcher argues that in many situations, as when prospective parents carry genes for genetic disease, cloning would be preferable to the genetic roulette of sexual reproduction (NBAC 1997, 40). Taking a view radically different from Roman Catholicism on what reproducing in a "human way" means, Fletcher submits that the deliberate, designed, chosen, and willed nature of laboratory-aided reproduction is distinctly human.

In his statement for the NBAC, Courtney Campbell (1998) reports on the views of Protestant theologians Paul Ramsey and Ted Peters. Ramsey gives voice to many Protestant denominations that warn we should not play God in our use of modern biotechnologies to create new life forms and replicate human beings. Peters counters by pointing out that we have a God-given responsibility to "play human"; that is, to use our creativity and freedom to chose a destiny of dignity for ourselves. Peters sees no theological reason that the alteration of human nature should be off limits to the creative powers of human intelligence, or that human cloning should be excluded from the toolkit used to chart our own destiny.

Rabbi Elliot Dorff (NBAC 1997, 46), who prepared a statement for the NBAC on the Jewish perspective on human cloning, wrote that humankind and God are cocreators, existing in an ongoing partnership to care for and improve upon creation. Support for this position is found in the Judeo-Christian creation myth when, in the first and second chapters of Genesis, the relationship between humans and nature is defined. Humankind is given "dominion over the fish of the sea, and over the fowl of the air, and over every living thing that moveth upon the earth. . . . And the Lord God took the man, and put him into the garden of Eden to dress it and to keep it" (Genesis 1:28, 2:15). The co-creator viewpoint argues that since humankind is given divine authority and the creativity to dress nature for its benefit, it should do so. Several Jewish thinkers approve of therapeutic cloning research that could save or improve existing lives. However, no religious tradition supports the notion that clones should be brought into existence solely for the benefit of a preexisting person.

Native American cultures. Speaking for Native American cultures, Courtney Campbell (1998, D-30-31) writes, "The whole of creation is good within Native American narratives, and all creation is animated, interrelated and responsible for harmonious interaction to sustain the order of life in the world. . . . Individual actions must be placed within a holistic perspective." Presuming this generalized worldview reflects most Native American cultures, Campbell supposes that most Native Americans would strongly oppose human cloning. Before his death in 2007, Gilbert Walking Bull, a Lakota spiritual leader and great-grandson of Sitting Bull, the Hunkpapa Lakota Sioux holy man, shared with me his thoughts about human cloning from his home in the Black Hills of South Dakota. His view is clear and uncomplicated: "Grandfather created everything different for a reason and it is not to be messed with or abused. Leave it alone. Respect Grandfather's creation. Respect for Grandfather and all the creation is central to Native American religion. Science should focus on strengthening the spiritual relationship to Grandfather within. No human can put a spirit in anything, only Grandfather can do that." If any segment of Native American culture would ever support the notion of human reproductive cloning, Campbell speculates that the support would derive from a belief in the preservation of endangered, indigenous people.

Buddhism. Faced with moral decisions about human cloning, Buddhists are not constrained by a worldview prescribed by a creation myth. Buddhism views time as running infinitely into the past and eternally toward the future. Central to teachings of the Buddha Gotama of the sixth century BC is awakening to the illusion of selfhood and a denial of the individual soul. Therefore, arguments against human cloning based on the wrongness of undermining a sense of "self" hold little sway in Buddhist culture. Campbell writes,

> Buddhist scholars generally agree that the process by which children are born into the world makes no difference. Cloning (can) thereby (be) understood as an alternative method of generating new human life. . . . The status of the human being is critical within Buddhist thought because it is the only condition by which an entity can achieve "enlightenment" and liberation from a world marked by suffering. . . . Some forms of Buddhism may endorse cloning because of the chance human life gives to achieve enlightenment. (1998, D-23)

On the other hand, most Buddhists would probably oppose human cloning if commercial or social agendas were placed above the best interests of the clone. Cloning research using sentient animals or human embryos would also be problematic due to the suffering and death caused by such experimentation.

Hinduism. On human embryo manipulation in the Hindu tradition, Campbell concludes,

> Classical Hinduism does not accept distinctions found in Western thought between God, human beings and other creatures, or between the supernatural, human nature, and nature. . . . Hinduism affirms a oneness of self with divinity rather than separation. A person cannot "play God" because in an ultimate sense the self is God. . . . The animating spirit is present from fertilization in classical Hindu thought . . . [and] the embryo is given the status of person throughout pregnancy. . . . The cultivation of spiritual self-awareness, rather than manipulation of the external environment, or one's biological self . . . is the overriding concern of the Hindu tradition. (1998, D-27)

Campbell writes that most scholars or practitioners of classical Hinduism would oppose cloning research on human embryos because of the embryo's personhood. However, he suggests that cloning for the purpose of relieving infertility or saving a life might be justifiable if it advances spiritual well-being.

Islam. Recall from chapter 2 that some Muslim traditions believe a soul enters the fetus only after four months of development. For these Muslims, cloning research on early human embryos would probably pose no moral dilemma. Cloning research might be viewed as a good thing if the ultimate objective is to promote health through the use of fetal tissue or ESCs to help repair worn out, diseased, or damaged organs. Islamic scholar Abdulaziz Sachedina suggested that his religious tradition could accept therapeutic uses of cloning "as long as the lineage of the child remains unblemished" (NBAC 1997, 54). However, any potential benefits of reproductive cloning would need to be weighed against its potential to disrupt and confuse familial relationships, which are highly valued on both moral and legal grounds within Islam.

Human Cloning Laws

As of 2012, no federal legislation regulates or prohibits human cloning in the United States, but several states have passed cloning legislation (National Council of State Legislatures 2008). Six states prohibit both reproductive and therapeutic cloning: Arkansas, Indiana, Iowa, Michigan, North Dakota, and South Dakota. Six other states prohibit reproductive cloning but allow therapeutic cloning: California, Connecticut, Maryland, Massachusetts, New Jersey, and Rhode Island. Arizona and Missouri do not ban reproductive cloning within their borders, but legislation prohibits using public monies for it.

Wisconsin's legislative history on cloning is interesting because the state is home to the first human ESCs created by James Thomson in 1998 (chapter 2). After lengthy and bitter debate that included much discussion in the public sector, Wisconsin's Assembly and Senate passed a bill outlawing both reproductive and therapeutic cloning and providing penalties of up to ten years in prison and $1 million in fines for violators. However, both votes lacked the two-thirds majority required to override a governor's veto, and in November 2005, Governor Jim Doyle vetoed the bill. In his veto message, Doyle cited the importance of human ESC research to the state's national image and biotech industry that generates $6.9 billion annually. Wisconsin's current governor, Republican Scott Walker, who survived a recall election on June 5, 2012, is known for his support of adult stem cell research over embryonic stem cell and therapeutic cloning research (Journal Sentinel PolitiFact).

Few countries have national laws addressing human cloning. Australia and the United Kingdom are exceptions. In Australia, a 2008 law prohibits thirteen specific procedures related to human cloning. Among these are "placing a human embryo clone in the human body or the body of an animal," "creating or developing a human embryo by fertilization that contains genetic material provided by more than two persons," and "developing a human embryo outside the body of a woman for more than 14 days" (Commonwealth of Australia 2008). In the United Kingdom, reproductive cloning has been prohibited since 2001. However, federally licensed therapeutic cloning is allowed. The 2001 legislation is described as "An Act to prohibit the placing in a woman of a human embryo which has been created otherwise than by fertilisation" (Department of Health 2001). Several international organizations, including the United Nations, the Council of Europe, and the World Health Organization, have produced statements or treaties to ban human reproductive cloning (Kendall 2009).[7]

Conclusion

Human reproductive cloning by SCNT seems imminently possible, but currently, physical risks to the clone preclude morally acceptable attempts to clone humans. Broad opposition to human reproductive cloning exists, and diverse objections to human cloning are raised. Objections based on religious belief ought to be respected, but they ought not necessarily be allowed to guide state and national policies. Few, if any, secular-based objections to human cloning appear to be valid reasons for completely prohibiting the technology. Nevertheless, several states in the United States and at least two countries have laws banning human reproductive cloning. For society to deal adequately with the possibility of human reproductive cloning, it must (1) acknowledge the need to educate people to the reality that all persons, cloned or not, have the capacity to live life with an open future, (2) address the problem of "personhood" in connection with human embryo research and the certain loss of embryos in research needed to make reproductive cloning safe, (3) assess the validity of the procreative liberty argument for allowing human cloning, and (4) address the problem of injustice if access to the technology is dictated by financial wealth.

Questions for Thought and Discussion

1. Are there reasons humans should not be cloned that were not mentioned in this chapter? If so, examine them critically for their validity.

2. Are there possible benefits of human reproductive cloning to individuals or society not mentioned in this chapter? If so, do the benefits outweigh possible objections? Can you counter the objections? Are the objections based on accurate knowledge about cloning?

3. Should human therapeutic or reproductive cloning be regulated or restricted by law? If so, should the laws be at an international level or at the federal level? In the United States, should human cloning regulation be left up to individual states? If you believe that international law should regulate human cloning, how would you enforce the regulations?

Notes

1. See http://www.geneticsandsociety.org/article.php?id=404#2003gallup (accessed June 22, 2009). As described on this site, "the Center for Genetics and Society is a nonprofit information and public affairs organization working to encourage responsible uses and effective societal governance of the new human genetic and reproductive technologies."

2. Byrne et al. 2007 is the technical report on therapeutic cloning in monkeys.

3. President Bill Clinton's letter is published at the beginning of the National Bioethics Advisory Commission's Report: "Cloning Human Beings." http://bioethics.georgetown.edu/nbac/pubs/cloning1/cloning.pdf (accessed January 20, 2012).

4. A clone and the cloned person would have different mitochondrial DNAs since mitochondrial DNA is inherited from the mother through the egg cell. The only exception to this would be if a woman had herself cloned and used her own egg cell for SCNT. Identical twins have long been assumed to have identical nuclear and mitochondrial DNA since they are derived from the same fertilized egg, but recent research shows that the nuclear DNA in identical twins may differ in the number of copies of certain base sequences, a phenomenon associated with disease resistance, autism, and lupus (Bruder et al. 2008).

5. NABER was an independent, private, not-for-profit board established in 1991 by the American College of Obstetricians and Gynecologists (ACOG) and the American Fertility Society (AFS) to review and comment on ethical issues related to new reproductive technologies. The board disbanded in 1998.

6. At the eight-cell stage, every cell of the embryo is totipotent, able to give rise to another embryo and, ultimately, to a normal individual, if it is isolated from the other cells of the embryo, cultured in vitro to the blastocyst stage, and transferred to the uterus. The procedure is called cloning by embryo splitting. It has not yet been applied to human embryos, but is commonly done with farm animal embryos.

7. See http://staff.lib.msu.edu/skendall/cloning/laws.htm (accessed July 6, 2009). On this site are links to detailed information on statements and treaties authored by international organizations. The site is created and maintained by Susan K. Kendall, Health Sciences librarian and liaison for the Biomedical Sciences at Michigan State University Libraries.

Sources for Additional Information

Campbell, C. S. 1998. "A Comparison of Religious Views on Cloning." In *Cloning: Science and Society,* edited by Gary E. McCuen, 119–39. Hudson, WI: Gary E. McCuen Publications.

Caufield, T. 2003. "Human Cloning Laws, Human Dignity and the Poverty of the Policy Making Dialogue." *BMC Medical Ethics* 4: 3. http://www.biomedcentral.com/1472-6939/4/3 (accessed June 10, 2012).

Emanuel, E. J. 1998. "Ethical Counterpoints on Human Cloning: An Overview."

In *Cloning: Science and Society,* edited by Gary E. McCuen, 71–76. Hudson, WI: Gary E. McCuen Publications.

Gibbs, N. 2001. "Baby, It's You, and You, and You . . ." *Time,* February 19, 46–57.

Gurdon, J. B., and V. Uehlinger. 1966. "'Fertile' Intestine Nuclei." *Nature* 210: 1240–41.

Human Genome Project. 2009. "Human Genome Project Information, Cloning Fact Sheet." http://www.ornl.gov/sci/techresources/Human_Genome/elsi /cloning.shtml (accessed July 6, 2009). This US Department of Energy site gives an excellent, nontechnical description of cloning and links to sites that describe cloning legislation in the United States and statements on cloning by diverse organizations.

Kass, L. 1997. "The Wisdom of Repugnance." *New Republic* 216 (22): 17–26.

Kolata, G. 1998. *Clone: The Road to Dolly and the Path Ahead.* New York: William Morrow and Co.

McCuen, G. E. 1998. *Cloning: Science and Society.* Hudson, WI: Gary E. McCuen Publications.

National Bioethics Advisory Commission. 1997. *Cloning Human Beings: Report and Recommendations.* Rockville, MD. http://bioethics.georgetown.edu/nbac/pubs .html (accessed July 4, 2009).

Nusbaum, M. C., and C. R. Sunstein. 1998. *Clones and Clones: Facts and Fantasies about Human Cloning.* New York: W. W. Norton & Co.

Pence, G. E. 1998. *Who's Afraid of Human Cloning?* New York: Rowman & Little-field Publications.

Polejaeva, I. A., S-H. Chen, T. D. Vaught, et al. 2000. "Cloned Pigs Produced by Nuclear Transfer from Adult Somatic Cells." *Nature* 407: 86–90.

Shiels, P. G., A. J. Kind, K. H. Campbell, et al. 1999. "Analysis of Telomere Length in Dolly, a Sheep Derived by Nuclear Transfer." *Cloning* 1 (2): 119–25.

Turner, L. 2004. "Is Repugnance Wise? Visceral Responses to Biotechnology." *Nature Biotechnology* 22: 269–70.

Wilmut, I., A. E. Schnieke, J. McWhir, A. J. Kind, and K. H. Campbell. 1997. "Viable Offspring Derived from Fetal and Adult Mammalian Cells." *Nature* 385 (6619): 810–13.

8
Age Retardation
Chasing Immortality for Better or Worse

> I argue thee that love is life. And life hath immortality.
> —Emily Dickinson, "That I Did Always Love"

One spring day, decades ago, my two best friends and I slowly walked the five blocks home from our grade school, where in a few weeks we would finish fifth grade. Sixth grade brought uncertainties for our friendship because we would enter junior high school, where there were three sixth-grade classes. We might become separated. Mulling over the realities of sixth grade, we promised each other and ourselves that we would live to be one hundred. For us, that was promising immortality so as to leave plenty of time for us to make it through junior high school and beyond, even in separate classes, and then afterwards to come back together, inseparable again. We did remain best friends through high school. But one of us did not live to see twenty, and another returned from Vietnam with shrapnel in his shoulder.

Nearly all of us wish to live a long, healthy life. Some may even harbor desires to live forever. Personally, I think it would be interesting to live for five hundred to a thousand years, just to see how certain human endeavors turn out. For instance, I would like to know whether we will communicate with extraterrestrial intelligence, our many environmental crises will be solved, humans will reach other planetary systems, we will abolish AIDs, the world's major religions will look the same, and war, terrorism, poverty, and prejudice will disappear.

Desires to live an extraordinarily long time seem most realizable when we are young. Children who are healthy and not traumatized by war, poverty, or famine feel virtual eternity stretching before them. I remember feeling that way. But, now mortality is a reality, and I am grateful for every day of health and life. What would it be like to live indefinitely, viewing the future like healthy, happy children do? This chapter probes humanity's endeavors to do just that, particularly its most recent scientific approaches to the project. We set out with these questions in mind:

1. Does death benefit the species?
2. What do the terms *aging, life span, life extension,* and *age retardation* mean?
3. How can we lengthen human life span?
4. What makes us age?
5. How can we retard the aging process?
6. How do some scientists believe we can make ourselves immortal?
7. How much life extension is likely in the foreseeable future?
8. What ethical issues accompany moderate increases in human life span?
9. What ethical issues accompany indefinite increases in human life span?

Age Retardation: The Biology

Natural selection gave us aging and death. Now science may be on the verge of thwarting the aging process, and someday it may even be able to prevent biological death. Our first consideration is to learn why we age and how death is beneficial.

Aging, Death, and Evolution

Since aging and death are so reluctant to yield to our efforts to abolish them, perhaps there are good, biological reasons for their existence and persistence. As is often the case for things biological, the reasons are tied to evolution. Natural selection acts upon traits that affect individuals' proficiency at successfully reproducing. The filter of natural selection can act only on individuals prior to and through their reproductive years. The mixing of genes during sexual reproduction provides the greatest source of genetic variation upon which natural selection acts. Any inherited gene combination or variant that helps an individual survive long enough to re-produce will be selected for and any variant that hinders survival through the reproductive period will be selected against. Natural selection tends to push the onset of diseases and infirmities that hinder reproduction be-yond the age of reproductive activity.

Consider Alzheimer's disease as an example. The disease is generally associated with old age, but even early-onset Alzheimer's strikes after the normal peak period for human reproduction, which is in the late teens and early twenties. Successful human reproduction requires intensive parental care for several years after babies are born. Therefore, conditions such as Alzheimer's disease work against successful reproduction, and natural se-lection tends to weed out genetic conditions that allow Alzheimer's to ap-pear before we pass our reproductive prime. On the other hand, genes that delay the onset of Alzheimer's disease until after the reproductive years are strongly selected for. These delaying genes predominate in all human popu-lations that experience Alzheimer's disease.

Multiply the Alzheimer's example by all of the diseases and infirmi-ties that accompany old age and you have a picture of the price we pay for sex. Actually, the price is paid by individuals who live long enough to ex-perience age-related infirmities. Our species and all other sexual species benefit from sex by the genetic variation it creates, which in turn provides evolutionary adaptability.

Only during the past century or so did we begin living long enough to experience what natural selection had pushed back beyond our repro-ductive years. Before then, childhood diseases, accidents, and harsh liv-ing conditions kept our lives so short that the ravages of old age were rarely seen. Now, due to improved childbirth procedures, immunizations, modern surgery, early disease diagnosis and treatment, refrigeration, safe water, and nutritious foods, persons in developed countries live well past their reproductive years. With longer lives come the unpleasant surprises that natural selection had hidden from us—cardiovascular disease, cancer, type 1 diabetes, arthritis, Alzheimer's and Parkinson's diseases.

So, what is the biological significance of death? It is to make way for the next generation containing a new pool of genetic variants. Without a continually renewed source of variants, natural selection cannot respond to challenges of new environmental conditions. Whether this applies to modern humans in a significant way is an open question. We have so cleverly insulated ourselves from environmental fluctuations that a vibrant source of genetic variation may not be as important for us now as it was ten thousand or fifty thousand years ago. For instance, we no longer need to generate individuals with extra thick layers of fat in order for the species to thrive at higher latitudes. Does modern technology make death less important to humans than it used to be? As we will see, there is no absolute answer to that question. But next we need a few well-defined terms in order to look at science's approach to the problem of human aging.

What Is Aging, and How Does Age Retardation Differ
from Life Extension and Life Prolongation?

Aging marks the passage of things through time. Material things change over time, and the accumulation of these changes is what we call aging. Everything ages except a beam of light.[1] Unless someday we move continuously at the speed of light, a highly unlikely happening, aging will always be with us. The question is to what degree we can control the changes that mark human aging.

Age retardation is slowing down the rate of undesirable changes to our minds and bodies as they move through time. Changes we may wish to retard include weakening muscles and bones, decreasing cardiopulmonary efficiency, wrinkling skin, memory loss, diminishing libidos and sexual functions, stiffening joints, and the onset of cardiovascular disease, diabetes, and cancers. We may wish to retain other changes accompanying aging such as the accumulation of wisdom through experience and the respect that sometimes comes with that. Age-retardation research and biotechnology aim to slow senescence of the human mind and body.

Life extension is increasing the period of active, productive, healthy, enjoyable living; and age retardation is one approach to accomplishing that. By contrast, life prolongation is simply protracting life, regardless of its quality. Medical technology is now quite good at life prolongation. That is why some people can exist for years or even decades in a comatose state. It is also why some persons give advance directives to refuse extraordinary means to sustain life or support legalizing physician-assisted suicide by terminally ill persons. This chapter does not address issues arising from life prolongation; rather, it focuses on the means and ethical implications of human life extension via age retardation.

Two more terms need defining. *Life span* is the duration of life for a particular individual, and *life expectancy* is the average number of years (or days) remaining in the life span of an individual at a specific point in time, assuming that the risk of dying at some future age remains unvarying for the remainder of her or his life. Life and health insurance companies use life-expectancy tables developed for persons with certain lifestyles or occupations (e.g., smoker versus nonsmoker; farmer versus teacher) and biologically based health risks (e.g., diabetes, obesity) to determine premiums.[2] A life-expectancy table for males in the United States indicates that this year my life expectancy is about 7,175 days and that I have almost a 99 percent chance of living to my next birthday. The same table shows that if I live to be ninety-nine, I will have a 64 percent chance of living to be one hundred.[3] Age retardation would increase the life-expectancy numbers in this table. How might that be done?

How Can Human Life Spans Be Lengthened?

Modern-day humans use three general approaches to lengthen life expectancies: (1) alleviate major causes of death among young and middle-aged people, (2) lengthen the lives of people who already live to the upper end of their life expectancies, and (3) slow the rate of onset of mental and physical infirmities that accompany aging.

The first approach includes efforts to reduce the number of childhood and teen deaths due to accidents and to reduce mortality in middle-aged persons due to heart disease, cancer, and diabetes. The second approach is to repair or prevent damage that afflicts organ systems during the natural process of aging. Strategies include using organ transplants, artificial organs, and stem cell therapy to replace dead or dying cells. This approach is like treading water in an ocean of aging processes, staving off the inevitable for as long as possible. The third approach is age retardation. The strategy is to learn about the underlying molecular causes for senescence and to devise means for halting or greatly retarding them.

Theories of Aging and Approaches to Age Retardation

The field of aging research is very dynamic. Top-notch molecular biologists are in hot pursuit of the molecular and genetic bases for aging, and some of them have even established biotech companies, hoping to capitalize on new discoveries that translate into age-retardation treatments. Some hypotheses for the cause of aging include preprogrammed cell death, genes for aging, the accumulation of mutations in the DNA of body cells, a buildup of cellular damage caused by *oxygen free radicals,* and hormonal signals for aging. Whether human aging is due mainly to one of these or a combination of factors is not yet known. Below, we delve a little deeper into some of the proposed causes of aging and also look at some interventions that have dramatic age-retardation effects in animals.

Are cells preprogrammed to die? Except for cancer cells and some stem cells, most cells grow and divide for only what appears to be a preset number of times. One theory of aging is that senescence and ultimate death are due to cessation of cell divisions in vital organ systems. Since cell regeneration is essential to maintain tissues throughout a lifetime, organs become weaker as division ceases in more and more cells. When dead cells are not replaced, organs simply wear out and death follows.

Just what do biologists now know about the natural limits imposed upon cell replication and the mechanism that enforces them? In 1961 Leonard Hayflick and a coworker at the Wistar Institute in Philadelphia showed that somatic cells cultured under laboratory conditions undergo only a certain number of cell divisions that seem preprogrammed according to animal species and cell type (Hayflick and Moorhead 1961). Hayflick worked primarily with human skin cells called fibroblasts that normally inhabit the spaces between tissues and produce components of connective tissue like collagen. These cells died predictably after about fifty cell divisions in laboratory culture, and the phenomenon became known as the Hayflick limit.

There are some cells that ignore the Hayflick limit and appear to continue dividing without ceasing. For example, cancer cells and some stem cells disobey the Hayflick limit. But most normal cells obey a Hayflick limit on their number of divisions. Nobody yet knows what limits the number of times a cell can divide. One possibility is the inexorable shortening of DNA molecules at the ends of chromosomes (telomere shortening). In chapter 7 we saw that beginning life with abnormally short telomeres may have led to premature aging in Dolly, the cloned lamb.

Hormones and aging. When levels of hormones such as estrogen and testosterone decline, physical and mental functions also decline. These observations led to hormone therapy to prevent bone loss by osteoporosis and to maintain sexual functions in postmenopausal women. Hormone replacement is uncommon in men, but a 1990 study showed that injections of human growth hormone reversed some aging symptoms in men aged sixty-one to eighty-one (Rudman et al. 1990). Effects of the therapy, administered over six months, included increased lean body mass, decreased fat tissue, increased lumbar-spine density, and increased systolic blood pressure and fasting glucose levels, but the researchers did not assess exercise endurance, muscle strength, or quality of life. A later study in men sixty-five to eighty-two years old showed that strength training significantly increases muscle strength and that growth hormone treatment did not increase muscle strength beyond that achieved by exercise alone (Taaffe et al. 1994).[4] Also, treating mice with a testosterone-related compound, dehydroepiandrosterone (DHEA), increased their life spans by as much as 40 percent.

Although hormone replacement/treatment does not retard aging in all tissues and organ systems, and growth hormone therapy may not be risk-free, some researchers believe that learning how hormones reverse some symptoms of aging will give clues to natural causes of aging.[5] Such information could lead to new approaches to age retardation.

Caloric restriction and longevity. As early as 1935, research at Cornell University showed that reducing calorie intake in laboratory rats by 30 to 40 percent increased their life spans by 20 to 75 percent. During the past two decades this phenomenon has been observed in diverse groups of animals, and its possible application to humans is appreciated.

Studies on organisms ranging from yeast, worms, spiders, and flies to mice, cows, dogs, and Rhesus monkeys show that restricting calorie intake by about 60 percent can increase life spans up to 50 percent. Even more exciting is that many of the animals in these experiments, including our monkey cousins, show diminished signs of aging, including muscle-mass retention and immune and nervous system functions characteristic of younger individuals. Since the work with monkeys began about twenty years ago and because monkeys live for several decades, the effect of calorie restriction on their life spans is not yet known. In 2002, a research project investigating the anti-aging effects of calorie restriction on non-obese humans began with funding from the National Institute of Aging.[6]

How decreasing food intake extends life spans is not known. But discovering why it happens may allow researchers to induce age-retardation processes by ways other than caloric restriction. This would be good news for persons who enjoy good food in ample quantities and the social aspects of fine dining.

Antioxidants and longevity. Our cells continually generate substances called *oxygen free radicals,* by-products of burned food molecules and other normal metabolic processes. Free radicals oxidize vital molecules like DNA and proteins, and oxidative damage can lead to cell death. Antioxidants like vitamins E and C neutralize the destructive effects of free radicals. We also obtain highly effective antioxidants by eating fruits and vegetables and drinking red grape juice and red wine. Unfortunately, even for persons with healthful diets supplemented by vitamin antioxidants, free radicals nearly always outnumber the antioxidants inside cells.

One hypothesis on aging is that cell death due to the relentless oxidative damage done by free radicals causes senescence and other symptoms of aging. This hypothesis is being tested by treating model organisms such as mice with synthetic antioxidants. Research on a chemical agent called WR-272 is an example. WR-272 was developed by the US military after World War II to protect soldiers against DNA damage caused by nuclear radiation, damage chemically identical to that caused by free radicals. Researchers now find that WR-272 injected into mice not only protects DNA from radiation damage but also makes the mice live longer. Whether the findings in mice hold true for humans and whether antioxidants like WR-272 truly retard the aging process remain to be discovered.

Genes that control aging. The genetic hypothesis for human aging is that the activity or inactivity of certain genes is what causes aging. In support of this idea is the remarkable discovery that mutating one gene in a group of select genes that seem to control the aging process in worms, flies, and mice produces life-extension effects even greater than caloric restriction. For example, a single gene alteration in roundworms increases life span by two to six times that of unaltered worms.[7] Mice have genes related to the age-retarding genes in worms, and mutating just one of these increases life spans of mice up to 50 percent.

If the genetic effects on aging in mice mirror what is possible in humans, people normally living to be 80 years old could easily be made to live for 120 to 160 years. But there could be unpleasant side effects to life extension via gene alteration. In the animals studied so far, these include reduced fertility, increased sensitivity to low temperatures, reduced size, and diminished ability to obtain a mate.

Humans as Spiritual Machines

Hormone therapy, increased antioxidant levels, caloric restriction, and gene alteration may each hold potential to increase human life spans to well past one hundred years while maintaining vigorous mental and physical health. But none of these approaches promises immortality or even life spans of several centuries. A few researchers have more grandiose visions for future human life spans. Some work to subvert the Hayflick limit on human cell division in hopes of finding a way to confer virtual immortality upon humans. Still others chase immortality by non-biological means.

In 2002 I visited with an internationally renowned computer scientist who was then in her or his mid-fifties. During our conversation, the scientist told me that she or he fully expects to become immortal via computer technology, whose power has increased exponentially for the past half century.[8]

The computer approach to human immortality presumes that the essence of our personhood lies in brain function, not in the appearance or functions of the rest of our body. If brain function gives rise to mind, and if our minds define who we really are, it is reasoned that by making our minds immortal, we can make ourselves immortal. A growing group of people believe that in this century it may be possible to literally download our minds, with all of their memories, emotions, and capabilities, onto or into material more permanent than the biological tissue it now inhabits. Once this is accomplished, so they say, it will not matter what type of body houses the mind. There might be completely artificial bodies to provide the mind with sensory information like our present bodies do. Computer scientist and artificial intelligence researcher Ray Kurzeil refers to humans at such a stage in their evolutionary history as spiritual machines. And novelist Greg Iles refers to a computer with a downloaded human mind as reaching Trinity state: "It's mind and machine, yet greater than both" (2003, 397).

Whether the twenty-first century brings human life spans of a few decades greater than what we have now or of thousands of years, it is not too early to begin thinking about the moral implications of effective age-retardation technologies. Let's try to imagine the impact of significantly longer life spans on individual and societal values.

Age Retardation: Human Values

Age-retardation technologies will have striking effects on individual lives and on the structure of society regardless of whether their use is limited or widespread and regardless of whether the technology adds decades, centuries, or millennia to present life expectancies. Our discussion of the effects of age-retardation technologies on human values focuses on four categories of concerns, those associated with (1) moderate life extension affecting the individual, (2) moderate life extension affecting society, (3) virtual immortality and the individual, and (4) immortality and society. Moderate life extension would allow life spans of 120 to 200 years, and by "virtual immortality" I mean life spans lasting 1,000 years or more. The human value issues raised by age-retardation technology are less about the technology itself and more about its effects on individual lives and the development of culture and society over time. In 2003, the President's Council on Bioethics (2003a) published a document on the moral challenges of applying age-retardation technologies to humans. Some ideas from this excellent essay are included in what follows.

Issues Arising from the Moderate Extension of Human Life Spans

The individual. Sometime during the first half of this century, it is likely that one or a combination of biological approaches to age retardation will extend human life spans to 120 to 150 years. How 150-year life spans will affect individual lives depends upon whether all stages of life are extended in proportion to their present lengths or whether just one phase of life, such as creative and productive adulthood, is extended.

Applying age-retarding technologies at or before birth to extend all stages of life raises the issue of autonomy for the new individual. Even if the procedure were reversible, the child could become thirty years old or older before developing the decision-making capacities now present in most eighteen year olds. By that time her or his untreated cohorts could already be well into professional careers and raising families.

How would a life in which all stages are extended by 50 to 100 percent differ from lives being lived today? Longer periods of childhood and adolescence would mean more years living at home with parents or guardians. Retarded rates of intellectual growth might mean taking at least 50 percent longer to move through academic material now taught during twelve years of public schooling. The same applies to higher education, and the added expense of more years at college and postgraduate education might deter many from educating themselves beyond high school. Imagine the personal relationship problems of partners with mismatched aging schedules. Persons could fall in love when they are in the same life stage but then find themselves in completely different generations after just a few years.

Spending 50 to 100 percent more years working would require several periods of retraining for persons in jobs with rapidly developing technologies. Longer periods of retirement could mean years of lower living standards unless retirement systems and personal savings plans were adjusted. Finally, the end period of life, if drawn out over twice as many years as is common now, could be correspondingly unattractive. These problems make it unlikely that humans will chose to retard aging during all stages of life. More likely is that age retardation will be applied to lengthen the most productive and creative period of life, and perhaps some persons will opt for age-acceleration treatments to shorten the end period of life.

How might young people be affected when they realize that their healthy, productive years may reach a century or more? We can imagine possibilities. Young persons are already noted for exuberant living that defies mortality. Positioning death even further into the future could encourage greater risk-taking and a lessened urgency to prepare for a career. Starting programs in higher education or technical training and leaving home might be delayed for several decades, especially when ample financial support comes from others. On the other hand, multiple careers during the prime of life would be feasible, but there may be a paucity of entry-level positions in many professions due to delayed retirements.

Family life and spiritual life might also be affected by 150-year life spans. Family reunions would be enormous, perhaps commonly including six generations. And for those whose spiritual life is motivated mainly by concerns over mortality, pursuing answers to life's persistent questions about meaning might be delayed for a century or so. Whether this would translate into less introspection during the prime of life is hard to say. Also difficult to predict is how spousal relationships might be affected by many more decades of prime-time living.

Society. How would moderate life extension affect national and global societies? Increases in the median age of populations by several decades mean either that many more persons will live long past retirement or that people will remain in the work force for decades beyond what is now usual. The former would stress social security systems and place financial burdens on workers to keep the systems propped up. The latter would keep younger generations from advancing to higher positions within their professions.

Increased longevity might be matched by decreased birth rates. This has already happened in developed countries experiencing increases in the median age of their citizens during the last century due to better health habits and care. Italy, Japan, and Spain already have birth rates below that required to sustain their present populations.

Lower birth rates could allay some of the upward mobility problems for persons early in their careers, but they also create new problems. Scientific, technological, and cultural innovations are primarily the creations of the younger generation. If carriers of old ideas live longer and bearers of new ideas become fewer, society could rapidly stagnate. The situation could create tension between a numerous, dominant subculture of older persons maintaining the status quo and a minority population of frustrated young people.

Do we have a duty to die? We might ask ourselves this question before we urge legislators to appropriate more funds for age-retardation research and buy stock in corporations engaged in this research.

Issues Arising from Indefinite Extension of Human Life Spans

The individual, religion, morality, and hope. Fear of death is a universal human trait. In fact, we may be the only species whose members know for most of their lives that someday they will die. The thought of nonexistence is discomforting for most of us, and humans use a variety of ways to gain solace while living with the certainty of biological death. Certain major religions promise immortality (Christianity and Islam) or reincarnation as an improved individual (Hinduism) if moral guidelines are followed or certain beliefs are adhered to. Religion plays an important role in societal cohesion and in the personal lives of believers. Religious doctrines and values furnish guides for behavior and decision making, impart purpose to living, and provide comfort in times of stress and sorrow. The Greek Prometheus myth identifies hope as the trait that distinguishes humans from other creatures. For billions of persons, a worldview provided by religion is the main source of that hope.

How would religion fare if humankind made itself immortal here on Earth? And if the religions that promise immortality or reincarnation disappeared, how would society fare? We can only guess at answers to these questions, but they deserve careful consideration as we apply science to our quest for immortality. One scenario is that people find "immortality religions" increasingly irrelevant as life extension becomes more successful. Some religions could vanish along with the specter of death. With assurance of biological immortality, incentives to adhere to moral guidelines promoted by religion might also vanish, leading to purposelessness, unbridled selfishness, and the ultimate breakdown of society. We will see shortly that one rebuttal to this scenario is the notion that fundamentals of moral behavior have an evolutionary basis.

A second scenario is that religions adapt to earthly human immortality as they have done before in response to discoveries and technologies emerging from the scientific revolution. For instance, Christianity no longer insists on Aristotle's Earth-centered universe, and most major Christian denominations have no quarrel with biological evolution.

A third scenario is a gradual fading of religious belief from human culture without the loss of moral guidelines for living. Consistent with this scenario is the contention of many sociobiologists and evolutionary biologists that human morality is a product of natural selection that promotes cooperation and successful social living. By this view, religion is not the ultimate source of moral principles. Rather, it simply reinforces behavior endowed upon us by natural selection.[9] Virtues like being loyal, protecting others, doing as you would have done unto you, causing no physical harm to other members of your group, respecting authority, having notions of purity and sanctity, and reciprocating both favors and misdeeds are all logical outcomes of the evolution of a social animal.

Let's return briefly to the specter of death. It is unrealistic to believe that biological death will ever be completely vanquished. There will still be accidents, murders, and organ malfunctions. Moreover, Earth will be engulfed by the sun in a mere four to five billion years. Even the unlikely possibility of immortals moving to other planetary systems to avoid incineration buys only a bit more cosmic time. Cosmologists tell us that our universe is expanding at a continuously accelerating rate that shows no prospect of lessening. So, the universe itself will someday succumb to a cold, black, entropic doom.[10] Since we are part of the universe, we must share its fate. Ever hopeful as it is, humanity may never fully conquer death, and beliefs offering transcendence from the material universe may remain with us for a long time.

Society and the species. Consider politics and human immortality. If a majority of persons opted for age-retardation measures that bestow millennia-long lifetimes, the birth rate would almost certainly decrease accordingly. Fewer new persons entering society means proportionately fewer new ideas, a situation that would favor the "old guard" remaining in control. The lack of innovation described earlier for a society with moderately increased life spans would be unimaginably magnified by indefinite life spans.

One can argue that long-lived individuals will acquire and create new ideas and find new solutions to long-standing problems. But imagine a candidate for US president with a three-thousand-year-long history of ardent conservatism who decides to champion a progressive program at odds with what she or he has stood for for three millennia. If today's candidates can barely live down their vote on a single bill a year ago, what are the odds that a person carrying thousands of years of political baggage associated with a particular stance on an issue will convince the electorate that she or he is genuine about a new position?

It is difficult to imagine vibrant evolution in art, literature, music, dance, and science if the same practitioners were to hang around indefinitely. The same is true for societal progress in areas of social justice. Would undeveloped countries remain undeveloped, or would the immortals use their indefinitely long lives to work for improved living conditions for everybody?

Finally, what about the species itself? If we made ourselves immortal and stopped reproducing, natural selection would cease and the species would lose its ability to evolve. Perhaps somatic cell genetic engineering would replace the genetic experiments of nature, and individuals could try out various combinations of genes in specific tissues during the course of their immortality. But, as we consider in chapter 10, replacing the process of evolution with societal fads and the marketplace may not be so wise.

*What Are the Prospects for Significant Human Age
Retardation in the Foreseeable Future?*

Writers on the subject of humanity's future differ in their assessment of the prospects for large-scale application of human age-retardation technologies. Science communicator Ben Bova (1998) reflects optimism in his book *Immortality*. He tells readers, "The first immortals are already living among us. You might be one of them" (1998). In contrast, age biodemographers S. Jay Olshansky and Bruce Carnes (2001) conclude their book, *The Quest for Immortality,* by advising us that hormone injections, antioxidants, meditation, or caloric restriction will not stop or reverse human aging. Instead, they offer recommendations for living out the potential life spans we were dealt at conception: "daily vigorous exercise (30–60 minutes per day); plenty of fruits, vegetables, and moderate amounts of low-fat protein; a restful sleep every night; an intellectually rewarding, nonstressful job, or no job at all; daily body massage; sex at least once a day; and a regular indulgence in your favorite vice: chocolate, barbecue ribs . . . at a rate of one or two per week for every decade lived" (Olshansky and Carnes 2001, 235).

My own view is that discovering the cellular mechanisms by which caloric restriction, antioxidant enhancement, gene alteration/addition, and hormone therapy retard aging will lead to treatments that extend human life by several decades. We ought to assume that this will happen within a decade or so and work now to solve the moral issues associated with moderate age retardation.

Conclusion

Age retardation, one approach to human life extension, slows or halts the undesirable effects of aging. Research in animals, from worms to monkeys, shows that caloric restriction, elevating antioxidant levels, genetic manipulations, and hormone therapy can retard aging. Whether these treatments act via a common mechanism is not yet known. Telomere shortening during DNA replication may limit how many times cells can divide before they die, and preventing this from happening is another approach to age retardation. Some computer scientists believe that non-biological-based immortality may someday be possible for humans by transferring brain function from mortal cells to immortal material analogous to silicon computer chips. Ethical issues at the individual and societal levels arise with the prospect of dramatic human age retardation. Individual issues include protecting the autonomy of the unborn or newly born who receive age-retarding treatments, distributing the technology justly, and decreasing career advancement opportunities for young people due to vigorous older people remaining longer in the workforce. Societal issues include adapting education, retirement, life insurance, and health care systems for populations with average life spans lengthened by several decades, if not centuries. Other likely effects on society are reduced birthrates, decreased sources for artistic, scientific, and political creativity and innovation, and exacerbation of existing societal disparities due to the selective use of age-retardation technology. If virtual, biological immortality becomes a realistic prospect for humans, we will need to examine its significance for the future evolution of our species.

Questions for Thought and Discussion

1. Do you favor using public funds for age-retardation research? Why or why not?
2. Find your life expectancy at http://www.ssa.gov/OACT/STATS/table4c6 .html or from a similar table. If you had the opportunity today to lengthen your life expectancy by any number of years, how much would you lengthen it?
3. Do you believe creativity in art and science would suffer if human life spans increased to 150 years and the birthrate halved worldwide? What if life spans were increased to several hundred years and the birthrate was lowered to maintain zero population growth?
4. Assume that within 50 years the majority of 20 year olds will have life

expectancies of 180 years, with at least 150 of those years being creative, vigorous, and productive. How do you think higher education and the college experience for those students would differ from that of current college students?

5. Would your present views on religion change if you were certain of biological immortality (or at least a life span of several thousand years) here on Earth? If so, how; if not, why not?

6. Should aging be treated as a disease to be cured?

7. Do you believe there are ethical issues raised by the prospect of human age-retardation technology not mentioned in this chapter?

Notes

1. Einstein's physics tells us that time ceases to pass for things traveling at the speed of light, and, so far, light is all we know that can travel that fast. The static nature of information carried in a beam of light is why the Hubble telescope can literally look back billions of years in time to when galaxies were first forming. Nothing has changed in the light beams carrying that information to us.

2. From the National Census of Fatal Occupational Injuries in 2006, published by the Bureau of Labor Statistics in 2007, the chances of losing your life on the job is thirty times higher if you are a farmer or forestry worker than if you are an educator. http://www.bls.gov/news.release/archives/cfoi_08092007.pdf (accessed May 19, 2012).

3. Life expectancies are in the Period of Life table published by Social Security Online and based on mortality rates in 2007. http://www.ssa.gov/OACT /STATS/table4c6.html (accessed September 12, 2009).

4. Eighteen men were used in this study (Taaffe et al. 1994).

5. Some data hint at a correlation between increased prostate cancer and elevated growth hormone levels.

6. The project is called CALERIE (Comprehensive Assessment of Long-Term Effects of Reducing Intake of Energy). http://calerie.dcri.duke.edu/index .html (accessed September 16, 2009).

7. The roundworm species *Caenorhabditis elegans* is the worm used in genetics of aging studies. It is a favorite experimental animal for developmental and cell

biologists because it is easy to grow in the laboratory, every cell division from the fertilized egg through the adult worm is mapped in time and place, and its entire genome containing about twenty thousand genes (roughly the same number as in humans) is sequenced.

8. This reflects a long-term trend in the improvement of computer hardware known as Moore's Law: the number of transistors that can be included in an integrated circuit, relatively inexpensively, doubles about every two years, allowing comparable rates of increase in processing speed and memory capacity.

9. Jonathan Haidt (2006), a moral psychologist at the University of Virginia, is a leading thinker in this area. He claims evidence for five pillars of moral behavior that natural selection gave to us during our evolution as social creatures: care, fairness, loyalty, respect, and purity.

10. Entropic doom is the logical fate of the universe due to the Second Law of Thermodynamics, which requires that every energy transformation results in increased disarray of the universe, provided that the temperature remains above absolute zero. The Third Law of Thermodynamics states that absolute zero is unattainable. A common misunderstanding of the Second Law is that processes like evolution and embryonic development, which show increasing order, disobey the Second Law. This is not true, since evolution and development and other order-producing activities like building construction occur at the expense of order in the universe at large. For example, highly ordered molecules in food and gasoline are burned in the process of constructing a cathedral or an individual, and the disarray created by burning to obtain energy is greater than the order achieved by the carpenters, masons, and the developing embryo.

Sources for Additional Information

About.com. 2009. "Why We Age—Theories and Effects of Aging." http://longevity.about.com/od/longevity101/a/why_we_age.htm (accessed September 16, 2009).

Angier, N. 2006. "Slow Is Beautiful." *Science Times. New York Times,* D1, December 12. This article tells about turtle longevity, physiology, and reproductive behavior.

Benecke, M. 1998. *The Dream of Eternal Life: Biomedicine Aging and Immortality.* New York: Columbia University Press.

Bova, Dr. B. 1998. *Immortality: How Science Is Extending Your Life Span—and Changing the World.* New York: Avon Books.

Hall, S. S. 2003. *Merchants of Immortality: Chasing the Dream of Human Life-Extension.* New York: Houghton Mifflin Company.

Heilbronn, L. K., and E. Ravussin. 2003. "Calorie Restriction and Aging: Review of the Literature and Implications for Studies in Humans." *American Journal of Clinical Nutrition* 78: 361–69.

Kurzweil, R. 1999. *The Age of Spiritual Machines.* New York: Penguin Putnam.

Olshansky, S. J., and B. A. Carnes. 2001. *The Quest for Immortality.* New York: W. W. Norton & Company.

President's Council on Bioethics. 2003. "Age-Retardation: Scientific Possibilities and Moral Challenges." http://www.bioethics.gov/background/age_retardation .html (accessed September 16, 2009).

Swaminathan, N. 2008. "Aging May Be Controlled by Brake and Accelerator Genes: Study in Worms Challenges the 'Rust' Theory of Senescence." *Scientific American,* July 24.

Vance, M. L. 2003. "Can Growth Hormone Prevent Aging?" *New England Journal of Medicine* 348: 779–80.

Wade, N. 2007. "Is 'Do unto Others' Written into Our Genes?" *Science Times. New York Times,* September 18.

West, M. D. 2003. *The Immortal Cell.* New York: Doubleday.

Yeoman, B. 2002. "Can Turtles Live Forever?" *Discover,* June. This article tells of turtle evolution, reproductive behavior, and telomerase activity.

9
The Mind

Neuroenhancement and Neuroethics

'Twas the week before finals and all through the dorm, few students were sleeping, since Adderall is the norm.

—Emily Gibson, M.D., "Barnstorming"

"A mind is a terrible thing to waste," wrote Professor Peter C. Doherty (2008), co-winner of the 1996 Nobel Prize for Physiology or Medicine, in an article about education in his native Australia.[1] Borrowing the slogan of the United Negro College Fund, Doherty (2008) asked of his fellow Australians, "Can we afford to waste a single, talented person?"[2] The human mind is hard to define. We know when someone has lost it, is wasting it, or is not in his or her "right" mind. And all of us agree that a strong, healthy mind is a good thing. But just what is the mind, and what does it mean to enhance the mind?

Neurophysiologist Antonio Damasio (1999, 337) writes that *mind* "refers to a *process,* not a thing. . . . What we know as mind . . . is a continuous flow of mental processes, many of which turn out to be logically interrelated." One dictionary definition of the mind of humans or other conscious beings is that which "reasons, thinks, feels, wills, perceives, judges, etc."[3] Modern-day scientists, physicians, and psychologists agree that the mind does not exist apart from the body, as seventeenth-century philosopher Rene Descartes thought it did. By contrast, the body can exist without a mind, as evidenced by late-term Alzheimer's disease victims and brain-dead patients. The mind depends upon a functioning brain and contains our personality, certain intellectual and behavioral talents and weaknesses, our creativity, and our sense of "self" moving through time. Domasio includes both conscious and unconscious elements in his concept of mind. For our discussion here, let's consider the human mind to be a product of the integration of conscious and unconscious cognitive and emotional elements of brain function that makes us recognizably unique persons to ourselves and others.

In the case of good minds, can there be too much of a good thing? By modifying certain aspects of brain function, we can enhance feelings of well-being and the ability to concentrate on intellectually demanding tasks? Future technologies will offer memory enhancement, selective memory loss, and other enhancements of the mind. These technologies will generate ethical dilemmas difficult to imagine today. Seven questions guide our exploration of biotechnology and the human mind:

1. What are neuroenhancement and neuroethics?
2. How do nerve cells work?
3. What are the main parts of a human brain?
4. How do *psychoactive drugs* work?
5. What new neuroenhancements are coming?
6. How could computers enhance brain function?
7. What are the benefits and objections to neuroenhancement?

What Are Neuroenhancement and Neuroethics?

Neuroenhancement is using drugs, machine-computer interfaces, or other technological interventions to augment or modify brain function. Of special interest are neuroenhancements that affect mood or cognitive aspects of brain function such as memory and attentiveness. Victims of brain injuries or disorders sometimes receive neuroenhancing treatments in efforts to restore normal brain function. These are therapeutic neuroenhancements. It can be hard to draw a line between therapeutic neuroenhancement and neuroenhancements aimed at elevating brain function beyond normal levels. Therefore, some bioethicists avoid the words *treatment* and *therapy* to describe any neuroenhancement. Instead, they use *enhancement* to describe any intervention that elevates the performance of a body component, including the brain.

In this chapter, I use *neuroenhancement* to describe non-therapeutic interventions simply to facilitate discussion of the ethical issues raised by boosting normal brain functions. A psychoactive drug is any chemical agent that affects brain function, regardless of its use for therapy or enhancement.

Neuroethics is a subdiscipline of ethics dealing with the manipulation of brain activity. During most of the twentieth century neuroethics dealt with moral issues related to the treatment of patients with brain or psychiatric disorders and brain injuries. In the twenty-first century, neuroethics addresses issues raised by new technologies for enhancing and imaging brain activities. In this chapter we examine the use of psychoactive drugs that affect mood and cognition and the use of brain-computer interfaces to augment brain function or operate electromechanical devices. New brain-imaging technologies also raise ethical issues, especially privacy, but are outside the range of this chapter.

Brain, Mind, and Neuroenhancement: The Biology

Our Two Nervous Systems

Humans and all other animals with backbones (vertebrates) have two types of nervous systems: a central nervous system (CNS) and a peripheral nervous system. The CNS consists of the brain, brain stem, and spinal cord. It receives, processes, and responds to information gathered by our senses. Our CNS allows us to speak, move, learn, remember, understand, experience happiness and desperation, plan, love, deceive, create, and sense our own consciousness. It is the target of neuroenhancers. The peripheral nervous system carries information between the CNS and other parts of the body like our sense organs and muscles. It allows us to see, feel, hear, taste, smell, and move.

During the past few decades, we have learned a lot about how our nervous systems work. Specific functions have been mapped to specific regions of the brain, and each week brings new information about how nerve cells store information and communicate with each other. But there remain vast chasms of emptiness in the body of knowledge about how physical attributes of the brain give rise to the elusive phenomena of mind and consciousness. Rather than speculate about these least-understood phenomena in human neurobiology, let's take a look at something that is quite well understood, the structure and function of a nerve cell.

Neurons and What They Do

Neuron is just another name for nerve cell. A neuron's shape reflects its function for receiving and transmitting information in the form of electrical signals. An archetypical neuron has three parts: a central cell body, a set of filamentous cell processes called *dendrites,* and an elongated *axon* that branches into several axon terminals at its end (fig. 9.1). At its end closest to the cell body, a neuron sends out hundreds of filamentous cell processes, called dendrites, to form a highly branched dendritic tree that receives signals from other neurons. At the other end, each neuron reaches out to send signals to other neurons (or to muscle cells) via a slender, elongated cell process called an axon. At its far end, each axon branches into several axon termini that relay signals to the dendrites of other neurons. Nerve cells conduct electrical signals in one direction—from dendrites, through the cell body, and outwardly down the axon. So the structure of a neuron is exquisitely specialized for signal input at one end and signal output at the other end. The numerous, tiny points of communication between neurons, where information is discharged and received, are called *synapses.*

A neuron is not only a signal conductor but also a mini-computer. A single neuron receives information from hundreds of other neurons through ten thousand or more synapses in its dendritic tree.[4] The many different signals entering a neuron may carry different messages. Some are stimulatory and others inhibitory, to various degrees. This cacophonous mixture of information flows through the cell's membrane to a specialized region of the cell called the axon hillock. There the diverse signals are interpreted and integrated. Then the neuron makes a mindless, biochemical decision on whether to propagate a signal down its axon to other neurons or to muscle cells. In the latter case, the muscle fiber must integrate signals from many different axons and "decide" whether to contract.

Chemical events happening at billions upon billions of synapses and

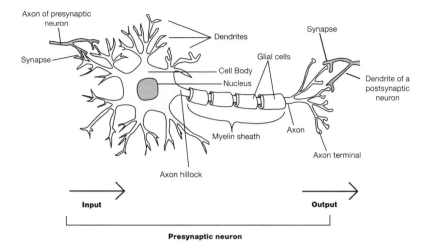

Fig. 9.1. A neuron with its three major parts: cell body, axon, and dendrites. Specialized glial cells comprise the myelin sheath that supports and insulates the axon and facilitates propagation of nerve impulses. Neurons communicate with other neurons at synapses. At a synapse, an axon terminal from the presynaptic neuron comes in close proximity to a dendrite from a postsynaptic neuron. Relative to the postsynaptic neuron (*far right*), the neuron depicted is presynaptic.

axon hillocks endow us with capacities to feel, smell, hear, see, taste, breathe, move, sense gravity and acceleration, react involuntarily, act deliberately, learn, remember, recognize, create, plan, experience, and do everything else that matters to us as animals and as persons. The title of a book by neurophysiologist Joseph LeDoux (2002), *Synaptic Self: How Our Brains Become Who We Are,* highlights the critical role of synapses and hillocks in forming our individuality.

Brain Structure and Humanness

Next we take a quick look at the one organ that most distinguishes us as a species, forms our aspirations, and mediates what we become, the human brain. To decipher how electrical communication between neurons is translated into self-conscious personhood is to crack what some neurophysiologists call the neural code. Understanding the neural code is one of the most interesting and most difficult challenges facing twenty-first-century biologists.

To fathom just how complex the neural code problem is, consider that the human brain contains about 100 billion neurons, roughly equal to all of the stars in the Milky Way, and that each neuron communicates with other neurons through at least one thousand synapses. This means that one brain operates with 100 trillion interconnections! If you were to try counting these synapses at a rate of three per second, you would be counting continually for over 10 million years. And the pattern of synaptic activity in the brain changes from instant to instant! Whether computer technology will ever be able to analyze, or even record, all of the interneuronal communications that go on inside one person's brain for even a few minutes is highly controversial.[5]

The externally visible brain in humans and other mammals has three distinct parts: the brain stem, the cerebellum, and the cerebrum (fig. 9.2). The brain component that most distinguishes humans from other mammals, including primates, is our extraordinarily large cerebrum (cerebral cortex). When a person is described as "cerebral," or not so, it is the cerebrum that is the object of praise or derision. Sometimes referred to simply as "the cortex," the cerebrum exists in two halves on the surface of the left and right hemispheres of the brain. The cortex is only a tenth of an inch thick, but if both halves of the cortex were flattened out, they would cover a square surface 4 feet and 3 inches on each side. This roughly equals the tops of two card tables and is four times the size of a chimpanzee's cortex. It is the front region of the cerebrum, the frontal lobe, that has expanded the most during human evolution. The frontal lobe seems to be the literal seat of our unique humanness, the place where neuronal activity produces self-awareness, plans for the future, and distinctly human thoughts.

Other regions of the forebrain, especially components of the limbic system, are evolutionarily older than the cerebrum. In fact, counterparts of our limbic system reside in modern-day reptiles whose ancestors gave rise to mammals over 250 million years ago.

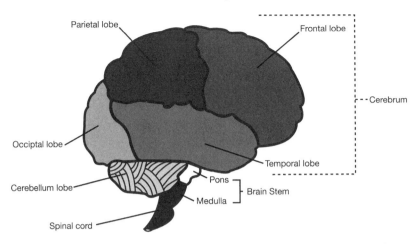

Fig. 9.2. The adult human brain with its three major parts: cerebrum (containing the frontal, parietal, occipital, and temporal lobes of the cerebral cortex), cerebellum, and brain stem.

Synapses and Neurotransmitters

Now we return to the synapse to discover why Professor LeDoux (2002) speaks of the "synaptic self." Recall that synapses are points of communication between signal-sending neurons and signal-receiving cells. The neuron transmitting the signal is called the *presynaptic cell,* and the neuron (or muscle cell) receiving the signal is the *postsynaptic cell.* The so-called *chemical synapse* between nerve cells is the most prevalent type of synapse in the CNS. Psychoactive drugs exert their influence at chemical synapses.

A chemical synapse is a tiny spot where two neurons almost touch and are in chemical communication with each other. It consists of very small patches of membrane in the corresponding neurons plus a narrow space between them called the synaptic cleft (fig. 9.3). Since nerve impulses cannot jump across the synaptic cleft, evolution endowed neurons with a way to convert nerve impulses into chemical signals that traverse the synaptic cleft.

Chemical communication between the two neurons happens when the presynaptic cell releases *neurotransmitter* molecules into the synaptic cleft

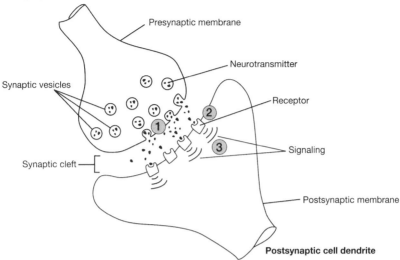

Fig. 9.3. A chemical synapse. In response to chemical and electrical signals, synaptic vesicles in the presynaptic cell fuse with the cell membrane and release neurotransmitter molecules into the synaptic cleft (1). Receptors in the postsynaptic cell membrane bind to the neurotransmitters (2) and send stimulatory or inhibitory signals into the postsynaptic cell (3).

and the postsynaptic cell detects them (fig. 9.3). The postsynaptic neuron's job is to process the many messages it receives and then "decide" how to respond in the form of a single signal sent (or not sent) down its axon to dendrites of other neurons. A look at events going on at a single synapse shows how neurotransmitters and some psychoactive drugs work.

Consider a synapse before, during, and after transmission of a chemical signal across the synaptic cleft. Before signal transmission, hundreds of small membrane-bound vesicles containing neurotransmitter molecules lay just inside each axon terminal of the presynaptic cell. These vesicles are called *synaptic vesicles,* and more than one type of neurotransmitter may be stored in the synaptic vesicles at a single axon terminus. In response to an electrical signal traveling down the axon, synaptic vesicles in the axon terminus fuse with the presynaptic membrane and release a rush of neurotransmitter molecules into the synaptic cleft. The neurotransmitters diffuse across the cleft and get grabbed up by *receptor* molecules embedded in the postsynaptic membrane (fig. 9.3). Properties of diverse types of receptors actually determine the nature of the signals received through a neuron's dendritic tree.

Axon terminals from over ten thousand different neurons may form synapses with a single postsynaptic neuron. By integrating signals from hundreds of thousands of receptor *molecules,* the postsynaptic cell makes a decision about whether to send an electrical signal down its axon to a thousand other neurons. Whether a postsynaptic neuron sends a signal down its axon depends upon the amounts and types of neurotransmitters released into the synaptic clefts of its dendritic tree, the numbers and types of its receptors, and how long neurotransmitters remain in the synaptic clefts.

Neurotransmitters disappear from the synaptic cleft in three ways: by re-uptake from the cleft back into the presynaptic cell, by degradation within the cleft, and by simple diffusion out of the cleft (fig. 9.4). As we will see, psychoactive drugs modify brain function mainly by affecting the levels and types of neurotransmitters or neurotransmitter receptors in the brain.

Psychoactive Drugs and How They Work

Most of us have methods to modulate our brain's performance. For alertness and a good mood, we get a good night's sleep, go to the gym, drink coffee, or inhale nicotine. To relax or escape unhappiness, we may drink alcohol. Such habits and activities can be considered neuroenhancers; so too are brand-name drugs prescribed for depression (Prozac), *ADHD* (Ritalin and Adderall), and social anxiety (Nardil). These drugs can also elevate mood or enhance mental performance in persons without diagnosed disorders. These and new generations of similar chemical enhancers that affect mood or cognitive abilities all act at the level of the chemical synapse. Knowing how some of these agents work gives insight into challenges faced by designers of new psychoactive drugs and an appreciation for the role of synapses in making us who we are.

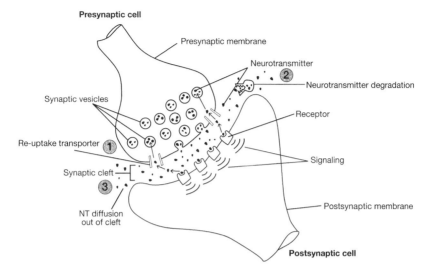

Fig. 9.4. Removal of neurotransmitter (NT) from the synaptic cleft. NT molecules are removed from the cleft by re-uptake transporters that return them to the presynaptic cell (1), by proteins that degrade them (2), and by simple diffusion out of the cleft (3).

Consider the action of fluoxetine, the generic name for Prozac. Fluoxetine belongs to a group of antidepressant drugs called *selective serotonin re-uptake inhibitors (SSRIs),* used to treat *major depressive disorder.* SSRIs selectively block the function of neurotransmitter transporters that remove serotonin from the synaptic cleft. Some SSRIs also block re-uptake transporters for dopamine and norepinephrine. The result of blocking re-uptake transporters is a prolonged presence of neurotransmitter molecules in the synaptic cleft and a resulting enhancement of their action (fig. 9.5).

Until about a decade ago, the prevailing hypothesis was that enhanced action of serotonin brought on by Prozac and other SSRI antidepressants, including Zoloft and Paxil, was responsible for the drugs' antidepressant activity. It was believed that serotonin (and perhaps also dopamine and norepinephrine) is a "happiness-inducing" neurotransmitter and that using SSRIs to keep neurotransmitters in the synaptic cleft longer made depressed persons happier.

Now new research shows that SSRI antidepressants may act by altering

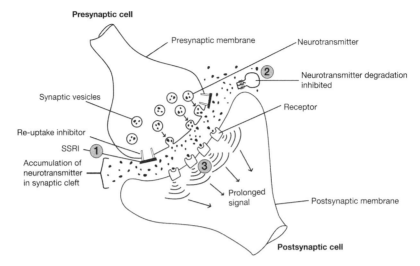

Fig. 9.5. Neuroenhancer action at chemical synapses. Some neuroenhancers act by blocking the re-uptake of neurotransmitters (NTs) by re-uptake transporters (1) or by inhibiting neurotransmitter degradation (2). The result is a prolonged signal caused by NTs binding to their receptors in the postsynaptic cell membrane (3). SSRI is selective serotonin re-uptake inhibitor.

gene expression in cells of the CNS to increase production of proteins called nerve growth factors (NGFs). Over time, NGFs act to increase the number of synapses in certain regions of the brain by stimulating neurons to grow more dendritic branches. A more highly branched dendritic tree allows a neuron to communicate with many more other neurons. Current thinking is that strengthening networks of neuronal communication in the brain makes for a happier person.

As a second example of psychoactive drug action on a synapse, consider methylphenidate and mixed amphetamine salts, the generic names for Ritalin and Adderall, respectively. Both drugs are CNS stimulants used to treat ADHD. Users of one or the other drug report that it helps them to mentally focus on their work and organize their lives. Both Ritalin and amphetamines like those in Adderall and Dexedrine increase the levels of dopamine and norepinephrine in synaptic clefts.

Finally, consider phenelzine, the generic name for Nardil. Its mechanism of action is different from the drugs described so far. Phenelzine is prescribed to treat anxiety disorder, some types of depression that do not respond to drugs like Prozac, and social phobia.[6] It belongs to a class of drugs called monoamine oxidase (MOA) inhibitors. MOA is a protein in the synaptic cleft of some synapses, and its job is to destroy neurotransmitters like serotonin, norepinephrine, and dopamine that have been released into the cleft. MOA inhibition causes an accumulation of neurotransmitters in the cleft and an enhanced signal to the receptors in the postsynaptic cell. The action of phenelzine is depicted by the inhibited neurotransmitter degradation agent in the synaptic cleft of figure 9.5.

Psychoactive drugs act at the synapse to elevate neurotransmitter levels in the synaptic cleft. Different types of agents accomplish this in different ways, and the varying effects of different drugs on mood and behavior reflect the types of receptors in postsynaptic cell membranes. But the synapse is where the action is. Ultimately, human nature itself resides in complex patterns of synaptic activity inside in the brain. Neuroscientist Judith Hooper writes dramatically about the potential of 100 trillion synapses in the human brain: "Hell, you can do anything with that. That's more than enough to contain a soul" (Hooper and Teresi 1987, 31).

New Psychoactive Drugs on the Horizon

Consider now what neurochemists have in mind for future enhancement of our "synaptic selves." Current psychoactive drug research focuses on two major goals: (1) to improve current mood-enhancing drug offerings by designing new drugs with lessened side effects and greater specificity of action, and (2) to develop new drugs that enhance or modify brain functions such as memory and learning for which no effective neuroenhancers yet exist.

Recent basic research unravels some long-standing mysteries about the sites in the brain responsible for positive mood and the biochemical bases for feelings of well-being. In his book *The Science of Happiness: Unlocking the Mysteries of Mood,* science journalist Stephen Braun (2000) writes about some of these discoveries and a "happiness set point" unique for each individual, around which mood fluctuates. Braun reviews the work on rats and humans in this area by researchers in Canada and at the Universities of Wisconsin and Colorado. Evidence indicates that happiness is an emergent property of biological properties of the brain, including epigenetic factors.[7] Ultimately, knowledge about brain function will allow pharmaceutics researchers to design drugs that can regulate a person's happiness set point in more nuanced ways than currently used antidepressants influence mood. Although the origin of individual happiness set points is not yet fully understood, it seems prudent to assume that in the foreseeable future pharmaceutical companies will be offering a new generation of mood enhancers that will allow persons to feel good most of the time.

Memory-enhancing and memory-erasing drugs are also on the near horizon.[8] Initially, memory enhancers will probably be used therapeutically for victims of Alzheimer's disease and non-pathological age-related memory loss. Eventually, they will undoubtedly become cosmetic pharmaceuticals used by anybody wishing to increase his or her memory abilities. An example of a memory-erasing drug is propanolol, an agent already used to treat high blood pressure and heart disease. Propanolol can also dull the memory of a traumatic event if taken within a few hours of the event. It and similar drugs will also probably be used therapeutically at first, for victims of rape and other violent crimes, for soldiers in order to ward off post-traumatic stress disorder, and for patients in certain emergency room situations. Propanolol does nothing to erase memories of long past events, such as abuse in childhood. But ongoing research aims to develop drugs that may ultimately make it possible to selectively erase memories from long ago.

Drugs to enhance learning are also coming. This became apparent nearly a decade ago when Tim Tully, a neuroscientist at Cold Spring Harbor Laboratory in New York, reviewed scientists' progress in this area (Tully et al. 2003). For example, Tully found that using certain drugs to increase the activity of a protein called CREB dramatically improves the learning abilities in learning-impaired mice.[9] Moreover, the drugs also reverse cognitive defects in mice with genetic disorders analogous to human Down syndrome, fragile X syndrome, and neurofibromatosis.[10]

There is ample precedent for the non-therapeutic use of neuroenhancers in the military and elsewhere. Dextroamphetamine (Dexedrine) has been used to combat wartime fatigue, especially by US Air Force pilots, in Vietnam, the Persian Gulf, the Balkans, Iraq, and Afghanistan (Bonné 2003; Woodring 2004; Emonson and Vanderbeek 1995). Pilots reportedly call the dextroamphetamine tablets issued to them "go-pills" and the zolpidem (Ambien) and temazepam (Restoril) pills issued to encourage sleep after a mission "no-go-pills." Various amphetamines are also commonly used by college students as study aids.

Brain-Machine Interfaces

Two fundamentally different classes of brain-machine interfaces exist. The first generates electrical signals that travel to the brain, usually via sensory nerve pathways. Cochlear implants and retina implants are examples of this type of device.[11] Signal-generating interfaces stimulate nerve cells in ways that the brain can understand in order to replace lost hearing, sight, or some other sensory function. In the second type of brain-machine interface, a receiver implanted in or near the brain actually picks up electrical signals from the brain and sends them to a computer for processing. The processed signals, derived from the subject's brain activity, are then used to operate some external device like another computer or a robotic limb. This type of brain-machine interface is called a brain-computer interface (BCI).

Experiments in animals and humans show that BCIs can do remarkable things. Famous animal BCI experiments were performed by researchers at Duke University with a macaque monkey that learned to control a mechanical, robotic arm with its thoughts. Related research by the same group resulted in a patent for an "apparatus for acquiring and transmitting neural signals" that can potentially be used to control weapons or weapons systems.[12] Another monkey experiment at the University of Pittsburgh School of Medicine showed that brain signals alone can operate a human-like robotic arm with enough precision to allow the monkey to feed itself marshmallows and pieces of fruit while its own arms are restrained.

In the medical arena, knowledge about brain function and BCI technology gained from experiments like these is aimed at restoring independence, mobility, or communication abilities to persons with paralysis, stroke, Lou Gehrig's disease, limb loss, muscular dystrophy, and brain disorders, including Alzheimer's disease, Parkinson's disease, and certain mental illnesses. Clinical trials with human subjects using a BCI technology named BrainGate are ongoing at Massachusetts General Hospital in collaboration with researchers at Brown University. So far, the human research is very promising. Subjects have moved computer cursors using neuronal signals from the motor cortex region of their brains.[13]

Neuroenhancement: Human Values

Neuroethics concerns itself with both the therapeutic and non-therapeutic use of psychoactive drugs and other interventions that affect brain function. We distinguish between therapeutic and non-therapeutic neuroenhancement here in order to facilitate discussion of concerns about the future use of neuroenhancements to elevate brain function beyond "normal."

The ethical issues associated with psychoactive drug therapy are not substantially different than those associated with other medical procedures; namely, the informed consent of clinical trial participants, safety (especially for children and adolescents), and distributive justice. By contrast, moral issues associated with the future non-therapeutic use of psychoactive drugs and computer-brain interfaces are relatively unexplored and have broad societal implications. Current use of unprescribed Ritalin, Adderall, and other amphetamine-containing pills by college students, entertainers, and others; ongoing advances in brain science; and the adeptness of pharmaceutical companies at marketing their products make neuroethics an important topic for public discussion.[14]

Arguments for Non-Therapeutic Neuroenhancement and Counterpoints

Three arguments for forging ahead to develop new, non-therapeutic neuroenhancements are (1) respect for personal freedom, (2) projected benefits to society, and (3) their inevitability.

The first argument goes like this: respecting the freedom to do with one's mind and body what one wishes to do, so long as the action is not illegal and does not cause injury to others, demands that the responsible use of neuroenhancements be allowed. The main difficulties with this argument are in determining whether others might be injured by one's use of neuroenhancements and in deciphering what is meant by *responsible* use of neuroenhancements. Not injuring others requires, at the very least, that the motivation for neuroenhancement be self-improvement and not the deliberate degradation or subjugation of others. But self motives can be difficult to examine objectively, and identifying others' motives is just as difficult. Individual, informed soul-searching, perhaps facilitated by counseling and education services, will be needed to help people assess their motives for neuroenhancement. This will be especially important in situations where neuroenhancement gives persons a competitive edge in obtaining jobs, gaining higher education opportunities, attracting a mate, or waging war. Responsible use of neuroenhancement may mean different decisions for different persons. At a minimum, responsible use of neuroenhancements will require prior self-education about their risks to self and others.

The second argument for neuroenhancement is that a better, happier world will result from increased numbers of contented people with better intellects. The argument is that persons with enhanced cognitive abilities will be more productive and creative, in turn benefiting the economy and enhancing standards of living for everyone. Likewise, it is presumed that the contentment of persons taking mood-enhancing drugs will have a trickle-down effect, improving relationships and elevating the level of happiness within society as a whole. What are we to make of this contentment argument? Drug-assisted mood stability already improves the lives of persons suffering from clinical depression, ADHD, schizophrenia, bipolar personality disorder, and other brain-based disorders. Why not improve everybody's mood on a regular basis? In 1932, Aldous Huxley warned us in his *Brave New World* that widespread, drug-induced contentment can compromise society's ability to recognize unhealthful living situations and make its members more vulnerable to exploitation.[15]

The third argument advocating neuroenhancement is that new neuroenhancement technology is inevitable and it is pointless to try preventing its coming by restricting its development or use. This is not a good argument for neuroenhancement per se. Responsible acceptance is not passive. Everybody has good reason to be informed about the prospects of future neuroenhancements and their effects on individuals and society. Responsible acceptance of the inevitability of neuroenhancements must include careful planning about how best to use them.

Arguments against Neuroenhancement and Counterpoints

Some bioethicists raise objections to non-therapeutic, chemical neuroenhancement (Fukuyama 2002; President's Council on Bioethics 2003; Parens 1998; Franjfurt 2006; Sandel 2007), and others critique these objections (Dees 2008; Bailey 2003). The objections and critiques also apply to neuroenhancement via computer-brain interfaces. Let's evaluate some of these objections as objectively as we can to see what, if any, human values might be at stake via the widespread use of neuroenhancements.

Objection 1. Neuroenhancers are unsafe and may permanently change the brain.

Is uncertainty about their safety a good reason to refrain from developing neuroenhancers? Safety is an issue with any new technology or medical procedure. As noted in the context of human cloning (chapter 7), prior to the birth of the first test-tube baby in 1978, many persons feared that the procedure would cause severe physical or mental abnormalities. Now, perfectly healthy babies conceived by in vitro fertilization (IVF) comprise about 1 percent of the live births in the United States each year.[16] Similar stories hold for new surgical procedures and therapies like open-heart surgery, joint and organ replacement, and radiation therapy. Always there are risks, especially when a procedure is in its early days.

Some argue that procedures like heart surgery and chemical neuroenhancement are fundamentally different since the latter involves modifying the brain, the very seat of our *selves*—our mind, intellect, and personality. The argument is that if neuroenhancing drugs permanently change the brain and alter one's personality, memory, or some other mental quality that identifies a unique person, they violate our personal identity. The counterpoint is that we already use nonchemical means to permanently alter our brains. Education, psychotherapy, learning to ride a bicycle, and athletic, military, or religious training are all brain altering. Although a few may see some harm in one or another of these modes of brain alteration, we as a society recognize that the capacity to have mind-altering experiences like technical training or a college education is part of our humanity. Besides, the effects of currently used neuroenhancers like Prozac and Ritalin are not permanent. Continued usage is required for the sought-after effects.

Relevant to the safety issue for cognitive enhancement are side effects from genetic and pharmaceutical memory enhancement observed in mice. Along with their improved memories and learning abilities for complex exercises, many of the treated mice show a lessened ability to solve relatively simple problems. It is possible that they remember so much that it interferes with normal behavior. Freelance writer Jonah Lehrer (2009) points out an analogy between the cognitively enhanced mice and a patient studied in the 1920s by a Russian neurologist. The patient was born with a memory so extraordinary that he could recite from memory Dante's entire *Divine Comedy* after a single reading. But accompanying this unusual mental ability was an inability to comprehend metaphors, presumably because his brain was so focused on literal particulars. Apparently, there is an optimum balance between remembering and forgetting. Too much of one or the other causes problems.

Wisdom requires caution in developing and using new neuroenhancing drugs. Regulators must be vigilant in overseeing pharmaceutical companies, where pressure to get products to market is high. On the other hand, the safety issue is not unique to neuroenhancers and, by itself, is not sufficient reason to reject the development and careful use of chemical or other types of neuroenhancers.

Objection 2. Neuroenhancing drugs cross a moral line between therapy and enhancement.

Is using psychoactive drugs or other neuroenhancing interventions for non-therapeutic purposes immoral? Objecting to the use of neuroenhancements for non-therapeutic purposes is based on two premises: (1) that therapy and enhancement are clearly distinguishable from each other, and (2) that enhancement of mental functions beyond normal levels is a bad thing. Both premises are vulnerable to criticism.

Some bioethicists make a distinction between therapy and enhancement; others believe it is impossible to make the distinction and use *enhancement* to describe all situations where biotechnology is used to improve human performance or traits.[17] As examples, consider two currently used psychoactive drugs. Methylphenidate hydrochloride (Ritalin) may be used by a person with diagnosed ADHD and also by a medical student studying for exams. Similarly, modafinil (Provigil, Alertec) may be used by an elderly person to stem memory loss and also by military pilots and infantry persons to enhance mental alertness while sleep deprived.

Who can say whether one's memory loss is a disorder or how severe the symptoms of ADHD must be to qualify for therapy? The *Diagnostic and Statistical Manual of Mental Disorders,* published by the American Psychiatric Association (2000), is the most reliable and objective guide for diagnosing mental disorders.[18] The multiple authors of the manual acknowledge the difficulty in drawing permanent, clear-cut lines between disorders and subclinical conditions. The manual includes "not otherwise specified" categories for most groups of disorders that allow clinicians personal discretion in making certain diagnoses.

If therapy is reserved to correct what is abnormal, we must define what is normal. Normalcy for a trait is generally defined either by statistical calculations or by applying subjective evaluations.[19] For example, population geneticists consider a trait normal if it is present in 5 percent or more of the population, while an inherited variant of the trait expressed at a frequency below 5 percent is deemed due to a mutation. Yet there is danger in defining mutants as abnormal since evolution by natural selection requires variation within populations, and the continual emergence of genetic mutations is a normal biological phenomenon.

Subjectively, one might consider the most desirable traits to be normal and departure from this subjective standard of normalcy as abnormal, disabled, or diseased. Dwarfism, congenital deafness, intersexuality, and very short and tall stature are conditions some consider to be abnormal or disabling. But persons with these conditions may not consider themselves disabled and are pleased to be who they are.

Diagnostic creep is a term used by bioethicist Richard Dees (2008) to describe the classification of a condition as a disease simply because there is an intervention available to ameliorate it. For example, when I was in grade school, children who were especially talkative and active were considered extroverts, but after the advent of Ritalin and Adderall, children with those same traits began being diagnosed with ADHD and given prescriptions to relieve their symptoms. Imagine *diagnostic creep* applied to cognitive functions where IQ measurements, memory tests, or other assessments of analytical skills are used to determine who is eligible for therapy with neuroenhancing drugs or electronic devices. The arbitrariness of defining cut-off points for cognition therapy and problems with cross-cultural intelligence testing create an extensive gray area between therapy and enhancement in the realm of mental function. This vast territory of ambiguity lends support to the notion that neuroenhancement should be available to all who wish to use it.

On the other hand, there are clear-cut examples at the ends of the spectrum of cognitive abilities, from developmentally disabled persons and dementia victims to intellectual geniuses. If neuroenhancements are used to therapeutically lift those at the lower end of the spectrum into a space where they can function more independently in society, why should neuroenhancements be denied to those whose cognitive abilities comprise the central or higher regions of a Bell curve? For those who believe that therapy is fine but enhancement is not, the answer to this question must be to deny the smart from becoming smarter through drugs or other interventions. But, to reasonably hold this position, one needs to explain why chemical- or electronic-based enhancement of mental function is fundamentally immoral. How is it different from using a tutor to learn calculus, language tapes to learn a foreign language, or coffee to sharpen one's alertness? We will see how some persons respond to this argument shortly in objection 5, that neuroenhancers undermine human authenticity.

Objection 3. Neuroenhancement to gain a competitive edge is self-defeating because others can easily gain the same relative advantage.

Would widespread neuroenhancement be pointless? After every round of neuroenhancement, there still would be persons with relatively low, middle, and high mental abilities. One would likely remain in the same group he or she inhabited at the start. Saying that neuroenhancement under such conditions would be pointless can be criticized on three points: (1) The objection ignores possible benefits to society of elevating the cognitive abilities of entire populations. (2) It is unlikely that everybody would choose to use neuroenhancements, even if they were affordable for all. Everybody now does not choose to read books, study a foreign language, or work brainteasers in their spare time in order to increase their intellectual acuity. So, undoubtedly, there would be a competitive advantage to be gained by those who did use neuroenhancements. (3) This objection is not really a moral objection. It simply states the view that neuroenhancements would have no practical benefit. To make it a moral objection, its proponents would need to show that increasing the cognitive abilities of everyone would be a bad thing, causing more harm than good. It is difficult to imagine how a more cognitively endowed public would harm society, especially in democracies where the nation's strength depends upon well-informed, critically thinking citizens.

Objection 4. Neuroenhancers undermine good character and personal responsibility.

Does non-therapeutic neuroenhancement undermine the character and responsibility that come from diligence and hard work toward life goals and virtuous behavior? If people could lead virtuous lives relatively effortlessly by taking pills or by computer-brain interfacing, might humanity eventually lose its ability to strive for excellence in living? Aside from the difficulty of defining just what *virtue* is, this objection presumes that neuroenhancements to induce or promote most types of desirable behavior are just around the corner.[20] This is not the case. The foreseeable future will most likely bring new generations of drugs similar to Ritalin and Prozac to facilitate concentration and feelings of well-being. There will also be drugs to enhance short-term memory and weaken memories of traumatic events. Many areas of human mental life will remain untouched by chemistry, such as the will to strive toward long-term goals, to sacrifice for friends and family, to stand up for one's beliefs, and to be truthful, trustworthy, loyal, helpful, friendly, courteous, kind, cheerful, thrifty, brave, clean, and appropriately reverent.[21] Still, caution in using neuroenhancement, especially for mood elevation, is appropriate. For example, the best action for one in an abusive living situation that is destroying his or her sense of well-being may be to leave or work to correct the situation rather than to override warning signs with pills.

Objection 5. Neuroenhancers violate human dignity, preventing people from living authentic lives.

This concern stems from the notion of an "ethics of authenticity"—treating life as a project that, when successfully carried out according to a plan formulated by conscientious introspection, gives life significance. An authentic life is summed up in the phrase "She is true to herself," or as Frank Sinatra crooned, "I did it my way." By contrast, an inauthentic life is one lived according to somebody else's plan or according to a plan formulated while one is not "himself." Some opponents of neuroenhancers fear that they will alter users' outlooks on life so that they are no longer themselves. This may lead to sidestepping one's true self when reflecting about how to live and, ultimately, to living an inauthentic life.

Countering this argument is the feeling reported by some users of Prozac that the drug actually allows them to finally feel like themselves (Kramer 1997). Who is to say what the true self is and whether somebody is in touch with it? Still, prospective users of neuroenhancers need to consider that a single drug is unlikely to be good at promoting all possible life projects.

Brain-Machine Interfaces: Special Ethical Concerns

Two major ethical concerns arise from brain-machine interfacing technology that is a decade or more in the future. The first is the potential for artificial sensory input technology to move beyond therapy and provide users with an alternative reality, not unlike in the film *The Matrix*. What will be the effects on individuals' relationships and on society as a whole if pornographic, murderous, and other exploitive experiences are easily downloadable into the brain and one feels the "experiences" as though they were real? What restrictions, if any, ought to apply to uses of artificial sensory input technologies?

The second concern is over BCI technology used to control mobile robots or other external devices with one's thoughts. How can we prevent use of such robots in criminal acts? How should legal systems treat crimes committed by robots? As it is, one cannot be tried and convicted for thinking about a criminal activity. But if one's thoughts were purposefully, or inadvertently, put into action by a robotic device, where does the responsibility lie? How will we prevent a hacker from gaining control of your robot with his mind? How should military and law enforcement applications of such technology be regulated? Roboethics, a new subdiscipline of ethics, may be needed to deal with questions like these.[22]

Conclusion

Neuroenhancers include psychoactive drugs, computer-brain interfaces, and other technological interventions that augment or modify brain function. *Neuroenhancement* here refers to the non-therapeutic use of neuroenhancers. Neuroethics is a subdiscipline of ethics that deals with the manipulation of brain activity. Neuroethics currently concerns itself mainly with ethical issues raised by new technologies aimed at enhancing, stimulating, or imaging brain activity. Personal identities are largely a function of brain activity, which, in turn, depends upon biochemical events at trillions of chemical synapses, points of communication between nerve cells. Chemical neuroenhancers act at the level of the synapse, usually to increase the amount or duration of neurotransmitter molecules in the synaptic cleft. Brand-name chemical neuroenhancers include Ritalin, Adderall, and Prozac. The latter and other antidepressants may also remodel the brain by expanding the dendritic trees of certain neurons. New generations of chemical neuroenhancers target mood, memory, and other cognitive functions. Computer-brain interfaces include therapeutic devices like cochlear implants that function to restore lost sensory function and sensors that detect electrical signals from brain cells and use them to operate external devices like prosthetic limbs. Ethical issues raised by neuroenhancers include safety, distributive justice, distinguishing between therapy and enhancement, social pressure for their use, and their effects on personal responsibility, authenticity, and societal norms.

Questions for Thought and Discussion

1. Should everybody be made as smart as possible?
2. Would it be desirable for everybody to be happy all of the time? Why or why not?
3. Should governments or health insurers see to it that everybody has an equal opportunity for mind enhancement? If not, how should we determine who is eligible for cognitive enhancements?
4. Do you believe there is a clear distinction between therapeutic and non-therapeutic neuroenhancement?
5. Should federal regulations restrict the types of neuroenhancements developed in the private sector?
6. Do you believe that future neuroenhancements will further stratify society into "haves" and "have-nots"? If not, why not? If so, is this a worrisome problem?
7. Should use of brain-machine interfaces or brain-imaging technologies be regulated?

8. What ethical issues not discussed in this chapter may arise due to the availability of non-therapeutic neuroenhancement?

Notes

1. Australian Nobel Laureate professor Peter Doherty writes in July 2008 about the educational system in his native Australia and how he thinks the intellectual talent of the country ought to be nurtured. http://www.sciencealert .com.au/opinions/20082705-17386-2.html (accessed July 30, 2008).

2. Since 1972 the slogan for the United Negro College Fund has been "The mind is a terrible thing to waste." http://www.adcouncil.org/default.aspx?id= 134 (accessed July 30, 2008).

3. This is the first of five definitions for mind found at http://dictionary .reference.com/browse/mind (accessed June 11, 2010).

4. Synapses may also form directly on the cell body or even along an axon, but they are most prevalent on dendrites.

5. Whether brain function can ever be mimicked by computers depends partly on whether the switch operation by a single transistor is roughly equivalent to activity across a single synapse. Professor Richard Chapman, Department of Computer Science, Auburn University, personal communication with author, 2009.

6. Social phobia is an anxiety disorder in which persons dread interacting with other people for fear of humiliating themselves, being judged cruelly by others, or being rejected.

7. A 2009 study of the epigenomes of cells from different areas of the brains of suicide victims, along with their case and family histories, suggests that abuse in childhood produces epigenetic changes to genes in cells of the hippocampus that may predispose these children to suicide later in life (Dolgin 2009). Dolgin (2009), a science journalist, reports on a 2009 technical report published in *Nature Neuroscience* 12: 342–48, by researchers at McGill University.

8. Ampakines and cyclic AMP response element binding protein (CREB) modulators are two of the former, and the drug propanolol exemplifies the latter.

9. CREB means cAMP response element binding protein. It is a transcription factor that activates the expression of genes for other proteins necessary for brain function, including the formation of long-term memories.

10. Neurofibromatosis is a genetic disorder characterized by benign tumors in nerve tissue and learning disabilities.

11. Cochlear implants work by processing sound waves into electrical signals that are transmitted to the auditory nerve by a device implanted in the inner ear. Retina implants work by processing visual images received by a camera into electrical signals transmitted to the optic nerve by an array of electrodes implanted on or behind the retina. Still under development, retinal implants are intended to restore some vision to persons with degenerative eye disease.

12. The research was supported by the Defense Advanced Research Projects Agency (DARPA), a US government agency responsible for developing innovative military technology. The approved patent application is at http://www.freepatentsonline.com/7187968.html (accessed September 2, 2009).

13. A new 2009 clinical trial with BrainGate and previous work on the project are described in a June 10, 2009, media relations release from Brown University: "Brain-Computer Interface, Developed at Brown, Begins New Clinical Trial." http://news.brown.edu/pressreleases/2009/06/braingate2?printable (accessed September 2, 2009).

14. In chapter 5, "Selling Happiness," of his book *The Science of Happiness: Unlocking the Mysteries of Mood,* Stephen Braun (2000) details specific examples of how pharmaceutical companies have altered the notion of "normal," attempted to suppress scientific data, selectively used statistics to analyze clinical studies in their favor to gain FDA approval for drug sale, and deliberately miseducated the public in order to market their products.

15. An interesting collection of quotations from Huxley's *Brave New World* about the nature and effects of soma is at http://www.huxley.net/soma/somaquote.html (accessed September 1, 2009).

16. See http://www.ivfbabies.com/ (accessed August 1, 2008).

17. Walters and Palmer (1997) distinguish between genetic therapy and genetic enhancement, describing the former as the prevention, treatment, or cure of disease. They suggest that disease is a departure from "species-typical functioning" and that "in each case a condition will need to be evaluated in the light of species-typical functioning and a judgment will have to be made about the extent to which the condition compromises such functioning" (Walters and Palmer 1997, 38–39. The Directorate General for Internal Policies (2009, 18–19) con-

siders "a treatment to be an enhancement if it is carried out to enable someone born with a bodily characteristic which is socially often considered a defect (such as cleft lip) to achieve a species-typical or nearly species-typical functioning." It proposes three categories of therapy and three categories of non-therapeutic enhancement. Authors who make no distinction between therapy and enhancement include Parens (1998) and Wolpe (2002).

18. The *Diagnostic and Statistical Manual of Mental Disorders* lists criteria for diagnosing nearly four hundred mental disorders. The fifth edition of the manual (DSM-V) is due to be published in May 2013.

19. Medical philosopher, ethicist, and author Lennart Nordenfelt (1995) makes this point in his *On the Nature of Health: An Action-Theory Approach.*

20. Plato devotes over thirty pages to the subject of virtue in his dialogue, "The Meno," without arriving at a definition for virtue.

21. This list of virtues is borrowed almost verbatim from the twelve points of the Boy Scout Law: to be trustworthy, loyal, helpful, friendly, courteous, kind, obedient, cheerful, thrifty, brave, clean, and reverent. I omitted "obedient" because I never did very well with that point!

22. A new book makes an important contribution in this area: Lin, Abney, and Bekey's (2012) *Robot Ethics: The Ethical and Social Implications of Robotics.*

Sources for Additional Information

Barondes, S. H. 2003. *Better Than Prozac: Creating the Next Generation of Psychiatric Drugs.* New York: Oxford University Press.

Baum, M. D. 2008. "Science & Technology: Monkey Uses Brain Power to Feed Itself with Robotic Arm." *PittChronicle,* June 9. http://www.chronicle.pitt.edu /?p=1478 (accessed September 2, 2009).

Billings, L. 2007. "Grappling with the Implications of an Artificially Intelligent Culture: Rise of Roboethics." *Seed Magazine,* July 16. http://seedmagazine.com /content/article/rise_of_roboethics/ (accessed June 12, 2012).

Braun, S. 2000. *The Science of Happiness: Unlocking the Mysteries of Mood.* New York: John Wiley & Sons.

Cowley, G. 2001. "Medicine: Searching for a New and Improved Prozac." *Newsweek,* December 31, 100.

Domasio, A. 1999. *The Feeling of What Happens.* New York: Harcourt.

Gilbert, S. F., A. L. Tyler, and E. J. Zackin. 2005. "What Is Normal?" In *Bioethics and the New Embryology: Springboards for Debate,* 215–26. Sunderland, MA: Sinauer Associates.

Glannon, W. 2007. *Bioethics and the Brain.* New York: Oxford University Press.

Huxley, A. 1932. *Brave New World.* New York: HarperCollins.

Kandel, E. R., J. H. Schwartz, and T. M. Jessell. 2000. *Principles of Neural Science.* 4th ed. New York: McGraw-Hill Companies.

LeDoux, J. 2002. *Synaptic Self: How Our Brains Become Who We Are.* New York: Penguin Putnam. An accessible book for nonscientists about the human brain and the cellular processes responsible for its function.

Lehrer, J. 2009. "Neuroscience: Small, furry . . . and smart." *Nature* 461: 862–64. This excellent article tells how research with mice gives clues as to how neuroscientists may be able to give human cognition a big boost, but not without drawbacks.

Levy, N. 2007. *Neuroethics.* Cambridge, UK: Cambridge University Press. An excellent overview of current issues in neuroethics.

Moss, H. 2003. "Implicit Selves: A Review of the Conference." Special issue: "The Self: From Soul to Brain," edited by J. LeDoux, J. Debiec, and H. Moss. *Annals of the New York Academy of Sciences* 1001: 1–30.

O'Shea, M. 2005. *The Brain: A Very Short Introduction.* New York: Oxford University Press. A great resource for the nonscientist.

Starr, C., and B. McMillan. 2010. "Nervous System." In *Human Biology,* 239–64. Belmont, CA: Brooks/Cole.

Whitehouse, P. J., and E. Juengst. 1997. "Enhancing Cognition in the Intellectually Intact." *Hastings Center Report* 27 (3): 14–22.

Wolpe, P. R. 2003. "Neuroethics." In *Encyclopedia of Bioethics,* 3rd ed., edited by S. G. Post, 1894–98. New York: Thomson Gale.

Synthetic Biology

From Cocreator to Creator

There is grandeur in this view of life. . . . [F]rom so simple a beginning endless forms most wonderful and most beautiful have been, and are being evolved.

—Charles Darwin, *On the Origin of Species by Means of Natural Selection*

"Better Things for Better Living . . . through Chemistry" (shortened to "Better Living through Chemistry" by the 1960s drug culture) was DuPont Chemical Company's twentieth-century slogan for almost fifty years. DuPont's "better living" claim refers to life with products like nylon, neoprene, Teflon, Mylar, Spandex, and Kevlar. Not one of these is found in nature, but each has certain advantages over cotton, wool, or hemp. If we haven't actually lived better due to DuPont's synthetic chemicals, most of us have at least enjoyed the convenience of using clothing, cookware, and car tires containing them.

The scientific and corporate worlds of the twenty-first century are now poised to offer another type of synthetics—new life forms created by *synthetic biology* (*synbio*). A decade from now we may hum to ourselves the tune to some biotech company's mutated version of DuPont's old slogan, something like "Superior Life via Synbio." What synthetic biology means scientifically and morally for us and the rest of life on Earth is an urgent question because synbio is on the move in university and corporate research laboratories around the world.

In this book's previous chapters I tried not to bias readers with my own views on how best to proceed with modern biotechnologies. Those chapters examined how biotechnologies can directly affect individual humans and human society. Synthetic biology is different from those previous technologies in that it will have long-reaching effects on virtually all of Earth's life forms. Therefore, I deal with synbio's ethical issues differently. I wear my biologist's hat and write explicitly about the preparations I believe we need to make for living with this biotechnology. We examine the following questions about the meaning and future of synthetic biology:

1. What is synthetic biology?
2. Does synthetic biology differ from genetic engineering?
3. How do scientists and engineers do synthetic biology?
4. How are new synthetic biologists trained?
5. What are synthetic biology's accomplishments so far?
6. What lies in the future for synthetic biology?
7. What ethical issues are shared by synthetic biology and other biotechnologies?
8. Is a special ethics needed to deal with issues raised by synthetic biology?
9. What projects should a synbio-ethics tackle?

Synthetic Biology: The Science

Synthetic biology is an interdisciplinary endeavor emerging from the convergence of molecular/cellular biology, *genetic engineering,* physics, chemistry, biochemistry, electrical engineering, *nanotechnology,* and information technology (fig. 10.1). Many researchers who call themselves synthetic biologists are actually trained in physics, electrical engineering, or computer science. Synthetic biology is a highly collaborative venture, drawing on the expertise, creativity, and goals of a mostly young cadre of scientists, engineers, and technicians. For convenience, we will use the term *synbiologists* for persons who do synthetic biology. In a way, we are all synbiologists because the products of synthetic biology will reflect the research we support and the commodities we wish for, purchase, and allow to be created.

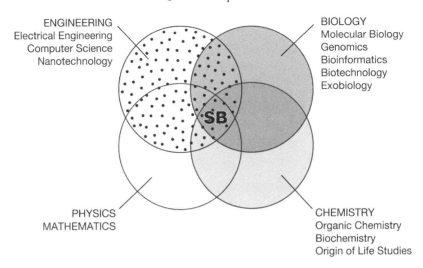

ENGINEERING
Electrical Engineering
Computer Science
Nanotechnology

BIOLOGY
Molecular Biology
Genomics
Bioinformatics
Biotechnology
Exobiology

SB

PHYSICS
MATHEMATICS

CHEMISTRY
Organic Chemistry
Biochemistry
Origin of Life Studies

Fig. 10.1. The interdisciplinary nature of synthetic biology (SB). Synthetic biology thrives on an amalgamation of ideas and methodologies derived from the life sciences, other natural sciences, engineering, and mathematics.

What Is Synthetic Biology?

What synthetic biology encompasses is summed up by the ETC Group in Canada, an international civil society organization "dedicated to the conservation and sustainable advancement of cultural and ecological diversity and human rights":[1] "Synthetic biology . . . [is] the design and construction of new biological parts, devices and systems that do not exist in the natural world and also the redesign of existing biological systems to perform specific tasks" (ETC Group 2007). Soon we will see what is meant by "biological parts and devices."

Synbiologists aim at nothing short of creating new forms of life to do humankind's biddings. In one sense, this is nothing new. Humans began creating new forms of plants and animals ten thousand years ago with the domestication of wheat, maize, dogs, and cattle through simple selection, hybridization, and breeding practices. In the 1970s, with the advent of genetic engineering technologies, we added bacteria to the list of organisms redesigned for our purposes (e.g., to produce human insulin and growth hormone); and since the 1980s we have transferred *genes* between species to produce transgenic plants and animals for agriculture, medical purposes, and new materials. Examples include corn and bean plants with bacterial genes for natural pesticides, rice with daffodil genes for better nutrition, pigs with human genes that make pig organs compatible for transplantation into humans, and goats with a spider gene for a protein that forms especially strong fibers, harvestable from the goats' milk. How does synthetic biology differ from genetic manipulations like these?

How Synthetic Biology Differs from Genetic Engineering

Some commentators on biotechnology claim that synthetic biology and genetic engineering are basically the same and that the regulations already in place for genetic engineering are sufficient for synthetic biology (Parens, Johnston, and Moses 2008).[2] Others refer to synthetic biology as "genetic engineering on steroids" or "extreme genetic engineering" to emphasize a qualitative difference between the two technologies (ETC Group 2007).[3] In this section we compare the two technologies, and the reader is invited to decide whether synthetic biology deserves to be in a scientific or ethical category by itself.

Genetic engineering and synthetic biology are similar in that both produce organisms with genomes different from the genomes made by nature. On the other hand, the two technologies differ in at least three ways: (1) the character of the life forms produced, (2) the underlying objectives of the technologies, and (3) the views of nature reflected and advanced by the technologies.

Life forms constructed by synthetic biology differ from those fashioned by genetic engineering. Genetic engineers tinker with the genetic material of organisms but do not fundamentally change the organisms themselves. Usually, the character of a genetically engineered organism differs from its non-engineered counterpart by just a single trait. For example, a corn plant given bacterial genes to render it resistant to insect pests is still a corn plant and still does corny things such as tassel, produce pollen, and develop ears of corn. By contrast, synthetic biology aims to create brand-new genes, gene circuits, *genomes,* and organisms with traits vastly different from anything invented by nature during life's 3.5 billion years on Earth.

Currently, synthetic biology focuses mainly on single-celled organisms (bacteria), but its vision for the future encompasses multicelled organisms. Stanford University synbiologist Drew Endy claims, "There is no technical barrier to synthesizing plants and animals, it will happen as soon as anyone pays for it" (quoted in ETC Group 2007, 7). Finally, synthetic biology's integration of designer genes into specially prepared genomes is much more exacting than gene transfer via genetic engineering, where the number and location(s) of foreign gene(s) in the host genome are difficult to control.

What about the underlying objectives of the two technologies; do they also differ? Genetic engineers work to soften irritating points of contact between us and nature. In a sense, they are cocreators with nature. They take what has evolved naturally and add to it or subtract from it in tiny ways at strategic points, leaving the core of the original life form intact. By contrast, synbiologists seek to create a new "nature" by bringing entirely new organism into existence.

Despite this difference between the objectives of the two technologies, that is, to cocreate with nature versus to create *de novo,* genetic engineers and synbiologists are rarely two separate groups of persons. Synthetic biology depends upon laboratory techniques in the genetic engineer's toolkit. The interdisciplinary nature of synthetic biology means that molecular biologists and geneticists may team with electrical engineers and computer scientists with a "Lego set components" approach to creating new life forms.

Finally, synthetic biology encourages a nonconventional view of biological nature. Ethicists Joachim Boldt and Oliver Müller (2008, 387–88) note, "Seen from the perspective of synthetic biology, nature is a blank space to be filled with whatever we wish."[4] They suggest that metaphors used in synthetic biology to describe DNA segments, combinations of segments, and new life forms may gradually erode society's respect for nature's creations. They point to metaphors such as Biobrick, Legobrick, switch, biological part, software, circuit, network, hardware, chassis, and cellular factory, warning that long-term and widespread use of such metaphors may discourage valuing of nature's life forms for their own sake.

In the second part of this chapter we will consider whether differences between genetic engineering and synthetic biology—the kinds of life forms created, their underlying objectives, and their differing views of nature—demand new ethical thinking for the era of synthetic biology.

How Synbiologists Do Synthetic Biology

There are two basic approaches to doing synthetic biology: top-down (fig. 10.2) and bottom-up (fig. 10.3). The top-down approach uses DNA segments or genes with known functions and arranges them in creative ways to design new life forms with properties useful to at least some humans. Top-down synbiologists insert extensive swaths of creatively combined pieces of DNA into existing cells. The inserted DNA may be a mixture of genes from several other organisms, or it may be completely artificial—envisaged by humans and made by a DNA-synthesizing machine. One active area of top-down research is developing host cells whose own genomes are pared down to facilitate acceptance of the new DNA and acquisition of correspondingly new cellular life styles.

Bottom-up synthetic biology aims to create new life forms from scratch using readily available chemicals. Bottom-up synthetic biology is not yet a reality, but if and when it arrives, both the genetic information (DNA or some new type of genetic information-carrying *molecule*) and its cellular containers will be wholly human-made. Both top-down and bottom-up synthetic biology make humans the creators of brand-new forms of life that are dramatically different from nature's products of biological evolution.

Examples of top-down synthetic biology. Assembling a large collection of "standard biological parts" (SBPs) is the project of a research group at Massachusetts Institute of Technology (MIT). An SBP is a well-defined segment of DNA with a known function in living cells. Some thirty-five hundred SBPs now stored in super-cold freezers comprise MIT's Registry of Standard Biological Parts and are available to synbiologists worldwide.[5] SBPs are categorized under headings that reflect collaborations between biologists, physicists, and engineers: e.g., measurement, reporters, inverters, protein generator, protein coding, regulatory, tags, signaling, composite devices, and terminators. The idea is to mix and match the parts

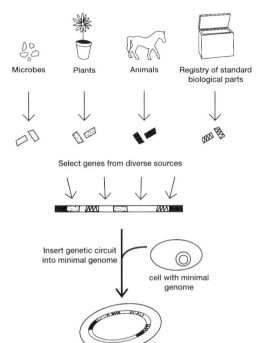

Top-Down Synbio

Microbes Plants Animals Registry of standard biological parts

Select genes from diverse sources

Insert genetic circuit into minimal genome

cell with minimal genome

Fig. 10.2. Top-down synthetic biology (synbio) uses genes or other components of existing organisms to generate radically new forms of life. Here "minimal genome" represents the bare essential genetic information needed to maintain microbial life. The minimal genome serves as a chassis upon which to add genetic elements from diverse sources that work together to direct the activities of a new life form.

to construct artificial gene circuits that function inside living cells and instruct them to do new things. It's like putting new software into a computer. Combinations of SBPs comprise the "software," cells themselves are the "hardware," and inserting the former into the latter commandeers cells to do the bidding of the software creators.

Likened to Lego bricks, SPBs are dubbed BioBricks.[6] Each BioBrick is customized to join to other BioBricks in certain conformations. Some Bio-Bricks are derived from living cells, while others are of purely human design. The ultimate aim is to assemble a collection of BioBricks large and diverse enough to allow the construction of almost any imaginable genetic outcome.

Other researchers, notably, Craig Venter of private and personal Human Genome Project fame (chapter 4; Venter et al. 2001), work to cus-

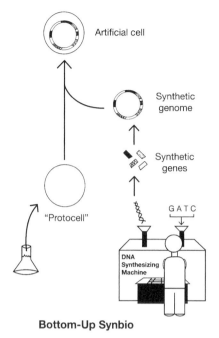

Bottom-Up Synbio

Fig. 10.3. Bottom-up synthetic biology (synbio) aims to build living cells from scratch. Here a human-made cellular container, the "protocell," receives human-designed genes made by a machine. The result is a hypothetical, artificial, living cell designed to perform a particular task or produce a specific product.

tomize "cellular hardware" to receive new "software"; that is, they prepare cells to absorb new genetic instructions and carry them out. To do this, Venter and others pare down the genomes of existing species of bacteria to the minimum number of genes required for life, basically leaving just the genes needed for metabolism and reproduction. For example, nature gave the genome of the bacterium *Mycoplasma genitalium* about 500 genes, but researchers (Glass et al. 2006) showed that the cell can live and reproduce with only 381 of these (fig. 10.4).[7] The researchers plan to transplant this "minimal genome" into some other bacterium to produce a new organism they call *Mycoplasma laboratorium*. This new species, and other microbes with minimal genomes, would be poised to receive synthetic gene circuits, perhaps constructed from BioBricks, instructing the cell to do wonderfully useful things or dreadfully destructive things.

Venter and his coworkers (Lartigue et al. 2007) at the J. Craig Venter Institute in Rockville, Maryland, showed the feasibility of their genome

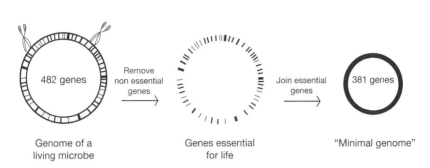

Fig. 10.4. Creation of a "minimal genome" from the natural genome of a living microbe. The natural genome of *Mycoplasma genitalium* contains 482 genes. Removing all genes not essential to the microbe's life leaves a minimal genome containing 381 genes absolutely essential to maintain life. The same procedure can generate minimal genomes from other species of microbes.

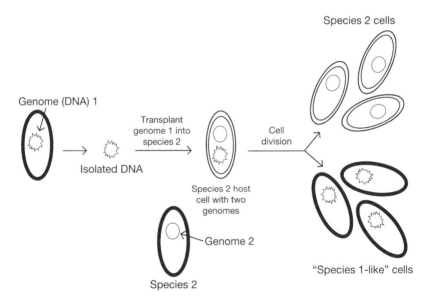

Fig. 10.5. Transformation of one species into another by genome (DNA) transplantation. Introducing the isolated genome from species 1 into a cell from species 2 containing its own genome produces a cell with two different genomes. When this cell divides, some of the daughter cells receive genome 1 and others receive genome 2. Daughter cells with donor genome 1 have the appearance and behavior of species 1 cells, as do all of the offspring of these cells.

transplantation approach in 2007 by moving the entire genome from one bacterium into cells of a different species of bacterium (fig. 10.5).[8] When the host cells containing two genomes divide, some of the new cells retain only the transplanted genome, and these cells acquire the characteristics of the genome donor cell. Thus, one species is literally transformed into another. Venter compared this accomplishment "to converting a Macintosh computer into a PC by inserting a new piece of software" (quoted in Than 2007). Venter's group (Smith et al. 2003) has synthesized complete viral and bacterial genomes from scratch, demonstrating the possibility of someday replacing a cell's genome with a genome designed and synthesized by humans from scratch.[9]

In 2010, Venter's group accomplished another major breakthrough for synthetic biology. The researchers (Gibson et al. 2010) synthesized the genome of one bacterial species, introduced the synthetic genome into cells of a second bacterial species, and got the synthetic genome to take over the host cell and work perfectly well.[10] The researchers nicknamed the living cell with a synthetic genome Synthia. To appreciate the significance of this work, recall from chapter 1 that a cell's genome is the entire DNA in the cell. The genome contains the genetic information needed for the cell to perform the activities necessary for life, including food acquisition and digestion, production of chemical energy, sensing environmental conditions and responding appropriately, and reproducing.

Venter's group deciphered the DNA base sequence for the entire genome of one species of bacterium, species 1. This sequence information was digitized for a computer and used by a *DNA sequence* manufacturer to direct the synthesis of DNA pieces, which Venter's group then spliced together to create the entire one-million-base genome of species 1 from scratch. Next, the workers introduced the synthetic species 1 genome into species 2 cells. In some of the descendants of the host (species 2) cells, the transplanted genome took over, replaced the natural genome of species 2, and actually converted the host cells into species 1 cells. The significance of the work is demonstrating that a completely human-made genome can direct life's activities, paving the way for future synthetic biologists to design brand-new genomes for life forms never before seen on Earth.

Venter's synthetic genome was not designed by humans, and he has not literally created life. The base sequence in the synthetic genome resulted from billions of years of evolution. Venter's group copied what nature had already designed and showed that their copy works as well as the natural genome. Embedded in the researchers' remarkable scientific achievement is also some scientific playfulness. Venter and his colleagues included four extra DNA base sequences within the synthetic genome that form four messages: one for decrypting the other three messages, one with the names of the researchers, one with three famous quotes, and the last with an email address. These molecular watermarks give the researchers a way to incontrovertibly distinguish between their synthetic genome and the natural genome of the host cell and thereby identify cells controlled by the synthetic genome. The watermarks also "brand" all cellular descendants of the researchers' work, which in the future may become distributed far and wide in other researchers' laboratories. Several controversial patent applications for the methods used to create Synthia are pending (e.g., US20070264688).

On May 20, 2010, the same day that Venter announced the results of this research, President Barack Obama directed his new Commission for the Study of Bioethical Issues to undertake a six-month study of the ethical implications of synthetic biology. We consider the recommendations of the commission later in the chapter. So far, Venter's genomic manipulations represent top-down synbio since the genomes they have engineered or synthesized already exist in nature. Let's see what some bottom-up synbiologists are doing.

Examples of bottom-up synthetic biology. An international team project funded by the Los Alamos National Laboratory, coordination center for the atomic bomb Manhattan Project during World War II, aims to synthesize life from off-the-shelf chemicals.[11] One objective of the research is to generate so-called protocells, cell-like membrane-bound spheres, into which human-designed DNA, RNA, or other types of genetic information-carrying molecules can be introduced. Nobel Laureate Jack Szostak at Harvard Medical

School performs similar work.[12] Szostak's group has created membrane-bound protocells containing encapsulated RNA, the biological molecule widely thought to be the precursor to DNA-based genetic systems in modern cells. The protocell vesicles grow and divide, distributing their encapsulated RNA molecules to the new daughter vesicles (Zhu and Szostak 2009). The simplicity and activities of these protocells suggest that similar entities may have formed on the prebiotic Earth.

One example of artificial genetic material is called AEGIS for Artificially Expanded Genetic Information System.[13] The goal of AEGIS work done by Steven Benner at the Foundation for Applied Molecular Evolution in Gainesville, Florida, is to expand DNA's four natural bases (A, C, T, and G) to twelve bases.[14] One possible outcome of AEGIS research is a genetic system that can code for proteins with more than the twenty different amino acids now used by nature. This would make it possible to build biological molecules and other structures with properties vastly different from anything yet invented by nature.

Some researchers doing bottom-up synthetic biology believe that humans are on the verge of actually creating life. Others believe that accomplishment is still many years off or may never happen. Whether wholly artificial life is ever created, there are spin-off benefits from the ongoing work of bottom-up synbiologists. The research gives insight into the nature of life itself and how the first life may have originated. It also provides clues to the forms life might take on extraterrestrial worlds, essential information for designing tests to detect alien life forms. Indeed, if we probe Jupiter's moons, Mars, and other extraterrestrial bodies only for Earth-like life, we may come up empty handed, missing one of the most profound discoveries humankind could ever make.

Future Synbiologists

Tremendous excitement about synthetic biology exists among a small but growing number of undergraduate students around the world. Since 2003, Tom Knight, creator of MIT's BioBrick registry, has run the International Genetically Engineered Machine (iGEM) competition. In the spring, student teams receive in the mail a collection of BioBricks, DNA segments dried onto filter paper or to the bottom of plastic sample wells. The teams' mission is to assemble the BioBricks in ingenious ways and get their genetic constructs to work inside living cells. In the fall, the teams converge at MIT for the annual iGEM Jamboree to show off their creations. Judging occurs and winners are selected. In 2008, eighty-four teams from twenty-one countries competed. The winning team from Slovenia engineered production of a vaccine against a bacterium that causes a common stomach illness.

In addition to learning synbio techniques firsthand and competing on a world stage with other young persons excited about synthetic biology, contestants submit their new genetic constructs to the Registry of Standard Biological Parts. The 2008 teams submitted fifteen hundred new Bio-Bricks, increasing the size of the registry by 75 percent.

Why do iGEM organizers target undergraduate students rather than advanced graduate students in molecular biology? *Scientist* magazine recently reported that "the iGEM competition is an unabashed recruitment tool, a scheme for seeding synthetic biology with young, energetic talent" (Katsnelson 2009, 44).[15] The iGEM organizers reason that graduate programs, perhaps unintentionally, begin instilling notions of what is and is not possible in their fields. Synbio promoters realize that for synthetic biology to accomplish spectacularly good things in the future, young synbiologists with unfettered imaginations must be nurtured. Preventing foolishly dangerous synthetic biology will require more than young, imaginative synbiologists though. It also requires wisdom coming at least in part from an appreciation for the evolutionary antiquity and functional complexity of the biosphere.

Some Early Products of Synthetic Biology

Only since 2002 has synthetic biology been recognized as a distinct discipline, but by 2007, over a dozen new patents were related to synbio. These include methods for DNA synthesis and protein design, several artificial BioBrick-like DNA components, a method for the complete chemical synthesis of genes and genomes, genetically modified bacteria that synthesize compounds useful for making certain pharmaceuticals, and a redesigned bacterium poised to receive new genetic instructions to convert it into a miniature, commercial factory for exotic molecules. In 2004 a team of undergraduate students at the iGEM Jamboree constructed a photosensitive bacterium. The students used a thin layer of the bacterial cells as a photographic film to produce a living, color photograph that spelled out the words *hello world,* a favorite phrase of computer scientists.[16]

In 2006 top-down synthetic biology struck a blow against malaria. Chemical engineer and synbiologist Jay Keasling and his team at Berkeley Lab and the University of California at Berkeley (Ro et al. 2006) redesigned yeast cells to produce artemisinic acid, precursor to a compound called artemisinin, the most effective anti-malarial drug known. Currently, artemisinin must be extracted from the sweet wormwood tree grown by small farmers in East Asia, Africa, and South America. Keasling and his group, backed financially by a $42.5-million grant from the Bill and Melinda Gates Foundation, soon hope to have microbe factories churning out low-cost artemisinin with the goal of saving the lives of nearly one million children per year. This sounds terrific, but we will see later that some unintended negative consequences accompany this project.

On the darker side, synbiologists have synthesized two deadly viruses. In 2002, researchers created a fully functional artificial poliovirus from mail-ordered DNA segments that they joined together in their laboratory. When it was injected into mice, the human-made virus caused paralysis and death. Then in 2005 other researchers at the US Armed Forces Institute of Pathology in Washington, D.C., Mount Sinai School of Medicine in New York, and the US Centers of Disease Control in Atlanta synthesized the genome of the Spanish flu virus that killed tens of millions of people worldwide in 1918 and 1919. Health officials thought this virus had disappeared for good when its last victims died decades ago. But then fragments of its genome turned up in frozen tissues of 1918 flu victims buried in Alaskan permafrost. Synbiologists used the genetic information in these DNA fragments to reconstruct the deadly virus from scratch. The purpose of the project was to examine the virus by modern methods of molecular biology to learn why it had been so deadly and then to design protection against similar viruses that might crop up in the future.

Notwithstanding the scientific and medical merits of the work, the re-creators of both the poliovirus and the 1918 flu virus were severely criticized for publishing their protocols in freely accessible scientific journals. Critics claim that reporting the research invites international bioterrorists to use similar methods to construct deadly pathogens. In reply, the researchers and some other scientists argue that reporting the work is justified on the basis of its value to disease prevention research and on the principle that science works best when it is transparent and unconstrained.

Future Applications of Synthetic Biology

The human imagination is the main determinant of future outcomes from synthetic biology. Craig Venter aims to construct bacteria that produce hydrogen from sunlight and water. Made from starting materials that exist in plentitude, hydrogen would rank with solar and wind power as a clean fuel. Venter predicts that "engineered cells and life forms [will be] relatively common within a decade" (Ferber 2004, 158). Other cell-size products coming down the pike from synthetic biology include programmable computers that use both analog and digital logic circuits, detectors and detoxifiers of environmental pollutants, underwater explosive sensors, monitors of human body health, and miniature, injectable, living machines programmed for gene repair or blood vessel rejuvenation. Synbio microbes may remediate climate change by removing carbon dioxide from the atmosphere, and cellular "factories" that churn out synthetic fibers for new generations of textiles and inexpensive pharmaceuticals are also on the horizon. Outcomes like these constitute the "Heaven scenario" for synergism among twenty-first-century genetic, robotic, information, and nano (GRIN) technologies described by Joel Garreau (2005), cultural revolution correspondent for the *Washington Post*.

Garreau (2005) also writes about a possible "Hell scenario." Intentional or unintentional actions of synbiologists could sponsor bioterrorism and threaten personal privacy and autonomy. In combination with the GRIN technologies, synthetic biology could create life forms tailored for surveillance, to cognitively impair, and to destructively target specific ethnic groups or individuals. Some commentators even go so far as to warn that misuse of biologically based technologies could spell the end of democracy (Fukuyama 2002).

As for synthetic biology's impact on the biosphere, nobody can know whether that will follow a hell or a heaven scenario. In a future when most of today's natural ecosystems may be known only from descriptions stored in digital databases, synbiologists may attempt to construct new ecosystems populated by synbio-created plants, animals, fungi, microbes, and other categories of life not now in existence. The purpose of these ecosystems will, of course, reflect human goals and desires. It is highly likely that pleasure, entertainment, the production of useful products that can turn a profit, and the self-satisfaction of simply filling ecological space with one's own creations will be among those goals and desires.

In summary, synthetic biology is an interdisciplinary endeavor to create new forms of life. It may benefit humankind tremendously in the form of improved health, new materials and energy sources, and environmental decontamination. With these benefits come risks of equal magnitude. In the second part of this chapter, we see that several of the ethical issues associated with synthetic biology are shared with other biotechnologies. Then we will ask whether synthetic biology raises any new ethical problems, especially in the context of humankind's emerging power as creator of life.

Synthetic Biology: Human Values

Synthetic Biology Raises Some Ethical Issues Raised by Other Modern Biotechnologies

Erik Parens, a bioethicist and senior research scholar at the Hastings Center, a nonpartisan research institution in Garrison, New York, dedicated to bioethics and the public interest, recently authored an article titled "Do We Need 'Synthetic Bioethics'?" (Parens, Johnston, and Moses 2008).[17] Parens and his coauthors claim that all of the bioethical issues raised by synthetic biology also exist for other technologies. They conclude that we just need to dig deeper into the old questions to find answers appropriate for the new technology of synthetic biology. Parens names four categories of ethical questions raised by synthetic biology and maintains that they have already been addressed for other technologies. The four categories of ethical issues are (1) safety and whether self-regulation or some form of public oversight is the best way to deal with risks, (2) how to deal with potential monopoly controls on various aspects of the technology and its applications, (3) how to justly distribute the benefits of synthetic biology, and (4) how far humans ought to go in reshaping themselves and other life forms. Let's look at each of these in the context of synthetic biology and consider whether to agree with Parens's view that all of these are "old" issues.

Safety, regulation, and responsibility. Unintended consequences of synthetic biology are a concern. For example, the release of new life forms, including multicelled organisms, into the environment could wreak havoc with natural ecosystems and even lead to species extinctions. Proteins generated from artificial genetic systems using amino acids foreign to nature's genetic system could cause allergic reactions. Some researchers claim that new life forms with artificial genetic systems are harmless even if they escape or are released into the environment because their genes cannot function inside cells with nature's traditional genetic system based on DNA bases A, G, T, and C (chapter 1). But this view ignores the possibility that synthetic organisms could drive native organisms to extinction through competition or upset the delicate balance of nature's ecosystems in other ways. The same holds for *genetically engineered organisms* for agriculture that must undergo extensive tests by the biotechnology companies responsible for their development and satisfy federal regulations before their release.

Currently, no special regulations govern synthetic biology, and the field is rapidly developing so that eventually persons with minimal training in biochemistry or molecular biology will be able to use readily available biological parts to construct new forms of life. Some commentators foresee a culture of "biohackers" similar to the computer hacker culture, whose creativity and entertainment is a daily threat to personal, business, and government computer systems (Dyson 2007).[18] Biohackers might compete to produce the most rapidly evolving, most bizarre, most artful, or most insidious life forms. Synthetic biology practiced by anybody at home in the garage is a frightening thought for some, while others predict that important discoveries could come from such unrestrained creativity. In any case, we must decide soon how synthetic biology can be regulated without overly restricting the scientific process and who should establish and enforce the regulations.

Another issue present for all new technologies is defining responsibilities for scientists whose work makes possible both beneficial and harmful applications. Some people claim that science is value free and that scientists bear no responsibility for the uses to which their research is put. However, most scientists and ethicists believe that the responsibility of scientists is to inform the public in accurate and comprehensible terms about the positive and negative implications of their research. Reciprocally, the public's responsibility is to require this of scientists. It is important to note that scientists are also part of the public. It is as important for a biologist to listen to a physicist explain the societal implications of nanotechnology as it is for an accountant or bartender to listen to a neuroscientist speak about cognitive enhancement.

Becoming informed and deciding on courses of action takes time, but advancing technologies don't wait for informed policies to develop. In the meantime, prudence may require following the so-called precautionary principle:

When an activity raises threats of harm to human health or the environment, precautionary measures should be taken even if some cause and effect relationships are not fully established scientifically. The process of applying the precautionary principle must be open, informed and democratic and must include potentially affected parties. It must also involve an examination of the full range of alternatives, including no action. In this context the proponent of an activity, rather than the public, should bear the burden of proof. (Martuzzi and Tickner 2004)

Patents and monopolies. Patents and monopolies pose ethical issues for virtually all of the biotechnologies we have discussed. Here are two examples from top-down synthetic biology.

In 2007 the J. Craig Venter Institute, Incorporated, created a furor when its namesake applied for US and international patents on an "artificial life form, a minimal life form version of the bacterium, *Mycoplasm genitalium,* including its genome."[19] If awarded, the patents would be the first legal claim on a species, and it would give Venter control of and income from its future use as a tool in synthetic biology. Venter is reported to say that he wishes to create "the first billion- or trillion-dollar organism" (Kaiser 2007, 1557). Some bioethicists discourage action on these patent applications until the societal and ethical implications of awarding them can be thoroughly examined. Action on the applications is still pending.

There is disagreement on whether patenting or an "open source" system is best for handling intellectual property rights for products of synthetic biology (Schmidt et al. 2008).[20] Some scientists believe that patents on tools for doing synthetic biology, like minimal genomes and synthetic genes, would slow advancement of the discipline. Others see patents as essential to ensure that discoveries and innovative laboratory methods remain accessible. Without patents, they argue, discoveries made in the private sector would remain company secrets. Still other scientists believe that an open source system similar to that used for the development, production, and distribution of computer software is preferable to patents (Open Source Initiative 2006). The open source argument maintains that improving synthetic biology's products and the rapid evolution of the discipline will occur best in the absence of patents. For example, an open source DNA system may not only require Venter to make his "minimal genome" version of *Mycoplasm genitalium* available to other scientists but also allow those scientists to modify it and pass the modified organisms on to others who may add their own modifications and so on. Obviously, any open source DNA system would need to respect the privacy of specific individuals' genetic information.

A second example of how synthetic biology patents and monopolies can have mixed positive and negative effects is the imminent large-scale production of the anti-malarial drug artemisinin by "factories" comprised of genetically redesigned yeast cells. To avoid emergence of artemisinin-resistant malarial parasites, artemisinin is used in combination with other anti-malarial drugs in a cocktail called artemisinin combination therapies (ACTs). The company Novartis has a virtual monopoly on ACT drugs because its particular ACT (Coartem) is the only one approved for purchase by United Nations agencies. This situation allows Novartis to use a two-tier pricing system for Coartem: low prices by contract with the World Health Organization for malaria-ridden countries in the Southern Hemisphere and much higher prices for international travelers and northern markets. Novartis's control of the ACT market and pending use of artemisinin produced by yeast cells threatens the livelihood of thousands of small farmers of wormwood, the natural source of artemisinin. Growing wormwood in the countries where artemisinin is actually needed to combat malaria, these farmers will be left without a market for their crop when Novartis switches to microbial factories for its source of the drug. The ETC Group suggests that "the Gates Foundation [guarantor of the synthetic artemisinin project] should insure that its focus on a synbio anti-malarial drug does not foreclose options for community-based, farmer-led approaches" (ETC Group 2007, 55).

Distributive justice. How can the benefits of synthetic biology be fairly distributed? This question applies to all new technologies and to the allocation of scarce resources such as human organs for transplantation and extraordinary medical care. Factors to consider include the sources of support for research and development that make possible the technologies, persons most in need of the technologies, who most merits the benefits of the technologies, and who can afford them. There are no easy answers to these questions. Fortunately, several of the likely near-term benefits from synthetic biology can be broadly distributed without much trouble. These include biofuels produced by microbial factories, a healthier environment from designer microbes that remove carbon dioxide from the air or inactivate harmful pollutants in the water and soil, and safer public places from explosive-detecting organisms. But as synthetic biology begins generating personalized medicine, enhancements for health or cognition, and other benefits unimaginable to us today, the distributive justice issue will become correspondingly more difficult.

The flip side of distributive justice is the question of who may be put at risk by certain outcomes of synthetic biology. Without proper regulations and safeguards, synthetic biology could yield organisms dangerous or deadly for the biosphere and all of humanity. Of course, unintended harm is also a risk with other biotechnologies, including genetic engineering and nanotechnology.

Reshaping life. Assisted reproduction and preimplantation/prenatal genetic diagnosis for gender or trait selection, neuroenhancement, surgical enhancement, and genetic engineering all involve the direct or indirect reshaping of living things. But among these, only germ-line genetic engineering reshapes life in a way that is passed on to future generations, a feature it shares with synthetic biology. Any argument that sets synthetic biology apart from germ-line genetic engineering as bringing up a brand-new ethical issue must show that these two technologies are fundamentally different. We examined such differences earlier and saw that synthetic biology differs from germ-line genetic engineering in three ways: the nature of the life forms generated, the technologies' underlying objectives, and the views of nature reflected and advanced by the technologies. These differences alone may not be enough to justify the development of a special ethics for synthetic biology, but additional factors may tip one's thinking in favor of a new subdiscipline of bioethics devoted to synthetic biology.

Is a Synthetic Biology Ethic Needed?

If one draws no qualitative distinctions between synthesizing brand-new genes, genomes, or cells (synthetic biology) and simply rearranging the genes that nature has already created (genetic engineering), synthetic biology raises no ethical issues not already addressed for the germ-line genetic engineering now used to produce agricultural and medical organisms with novel genes.[21] Such issues include safety, risk, patents, preventing monopolies, and distributive justice. This is basically the view of bioethicist Eric Parens and his coauthors, described earlier. Parens encourages developing an ethic applicable for all emerging technologies rather than worrying about the subcategories of neuro-ethics, nano-ethics, synbio-ethics, and so on.

A similar view is expressed by the Presidential Commission for the Study of Bioethical Issues in its December 2010 report on synthetic biology. The commission recommends a policy of ongoing "prudent vigilance" to monitor, identify, and mitigate risks and potential harms from synthetic biology. Among the commission's eighteen specific recommendations is establishment of a coordinating body or mechanism to regularly review developments in synthetic biology, assess the appropriateness of regulations relevant to the field, interact with the international community, and keep the public informed about its work. Regarding responsibility and accountability for safety and risk management, the commission (Presidential Commission for the Study of Bioethical Issues 2010) emphasizes corporate responsibility and self-regulation by the research community, application of the existing "NIH Guidelines for Research Involving Recombinant DNA Molecules" (National Institute of Health 2012), and adherence to the National Environmental Policy Act (Federal Emergency Management Agency 2010) for the field release of research organisms.[22]

Other knowledgeable and responsible persons take a very different view and argue that "prudent vigilance" is not an appropriate response to synthetic biology. A letter (ETC Group 2010) undersigned by fifty-eight organizations from twenty-two countries and written to Amy Gutmann, chair of the Presidential Commission for the Study of Bioethical Issues, maintains that the precautionary principle, not "prudent vigilance," should guide synthetic biology regulations. One component of the precautionary principle is that it shifts the burden of proof regarding the safety of an action to the proponents of the activity and away from those who are concerned about safety (Kriebel et al. 2001).

For those who believe the aims and practice of synthetic biology and genetic engineering are fundamentally different, the unique ethical problem posed by synthetic biology pits evolution against human hubris. The question is whether judgments of the marketplace and human desire can possess the "wisdom" of 3.8 billion years of biological evolution (Bradley 2010). Natural, evolutionary mechanisms create stable, interdependent networks of organisms in communities, ecosystems, and the biosphere as a whole. The central ethical problem for synthetic biology is to discern the special responsibilities that accompany our emerging role as creators of new life forms.

Four factors could combine, in the context of synthetic biology, to jeopardize the future health of life on Earth: (1) commodification of nature, (2) human exemptionalism, (3) trained incapacities in the disciplines of evolution and ecology, and (4) the sixth extinction now underway in our biosphere. Let's briefly consider each factor and how, together, they argue for development of an ethic specially tailored to guide future synthetic biology.

Commodification of nature. Commodification of nature refers to the mindset that values land mainly for its economic significance. In his famous essay *The Land Ethic,* Wisconsin naturalist and conservationist Aldo Leopold (1949) wrote against viewing land simply in terms of its economic value. Land is more than the board feet of lumber it can produce, its agricultural value, or its commercial worth for tourism, maintained Leopold, to whom "land" is all of nature, not just soil. Treating nature as a commodity to be exploited for corporate profit or human pleasure is the "key log" that must be removed before an effective "land ethic" can be developed, wrote Leopold in 1949. Six decades later a logjam of ignorance about evolution and ecology and of shortsighted, profit-centered attitudes still thwarts the development of a land ethic that recognizes the interdependency of all of nature's components.

In his book *The Creation,* Harvard University biologist E. O. Wilson (2006) identifies habitat loss, pollution, human overpopulation, and overharvesting as four of five human-caused factors responsible for the current dramatic decline in Earth's *biodiversity.*[23] Each stems directly from humankind's commodification of nature. The ETC Group offered an extreme example of commodification-based thinking by one synthetic biologist who suggests redesigning tree seeds so that they grow into houses rather than trees.[24] Synbio literature is replete with examples of life viewed as a commodity to be constructed for human purposes. With proper controls, constructing new life forms for human benefit is not necessarily a bad thing. But in combination with the additional three factors discussed next, the commodification of nature becomes an especially dangerous threat to life on Earth.

Human exemptionalism. Human exemptionalism refers to the philosophy that supposes humans can thrive outside the laws of nature. The concept was formalized into a Western philosophical/sociological paradigm soon after the eighteenth-century Enlightenment. Exemptionalist thinking held sway through the Industrial Revolution and well into the mid-twentieth century. Human exemptionalism is now severely critiqued by environmental sociologists and ecologists. Nevertheless, much of the general public and governance structure in the United States and other industrialized nations still live, vote, and behave as though humankind can live independently from the natural world. Characteristics of exemptionalist thinking include lack of interest in and denial of the role human activity plays in the current wave of extinctions and climate change. The attitude that mass extinction and loss of forests and other natural habitats do not affect human welfare and the view that preserving biodiversity is less urgent than economic growth, democratic expansion, military defense, or developing cures for cancers epitomize human exemptionalism.

Trained incapacities in evolution and ecology. The concept of *trained incapacity* is credited to Norwegian American economist and sociologist Thorstein Veblen (1857–1929), who wrote of business transactions being carried out with thoughts only of monetary gain and with little regard for their effect on the welfare of the community, thereby putting workers, the community, and the business people at cross purposes (1914). The implication is that business persons are trained to act with one goal in mind, corporate profit, and that this training produces the incapacity to see more broadly the implications of their activities. The concept has also been applied in sociology in the context of the eugenic potential of human genetic engineering (Duster 2003).

The trained incapacity concept is also germane to understanding the underlying causes of our environmental crisis. Failure to appreciate basic principles of biological evolution and ecology does not foster wise environmental stewardship. When stewardship of the biosphere includes decisions about generating and releasing new life forms into the biosphere, trained incapacities in evolution and ecology become especially worrisome because evolution describes how the biosphere came to be and ecology describes how it works. If synthetic biology, guided in part by human desires for profit and entertainment, supplants biological evolution and moves to replace disappearing natural biological communities with collections of designer organisms, the end result for life on Earth is apt to be counter to our likings.

That education in the United States fosters trained incapacities in evolution and ecology is irrefutable. A Gallop poll released on the eve of Darwin's two hundredth birthday in February 2009 reported that only 39 percent of Americans "believe in the theory of evolution" (Newport 2009). Biological evolution is often not discussed in high school biology classes in order to avoid conflict with religious fundamentalists who favor scriptural accounts of creation to the overwhelming evidence for the common ancestry of all living things. At colleges and universities, only some life sciences curricula require courses in evolution and ecology, and non-life science curricula require none at all. The upshot is that future voters, legislators, business persons, lawyers, lawmakers, CEOs, agriculturalists, medical professionals, physical scientists, engineers, and synthetic biologists receive a trained incapacity to think wisely about the future of our species and the rest of nature. They are deprived of understanding how the evolutionary origins of biodiversity relate to the structure and functioning of today's ecosystems and of comprehending humankind's absolute dependence on nature.

Although discussing solutions to this situation is off-topic for this book, science education initiatives by the American Association for the Advancement of Science (2011; also Caven 1992) and by the US National Academy of Science (2010) are off to a good start.

The sixth extinction. The final factor acting in combination with the previous three factors to give legitimacy to developing a synthetic biology ethic is the current and spectacular rate of species extinctions, the sixth extinction.[25] Environmental biologists estimate that known and unknown species of plants, animals, and microbes are now disappearing at the rate of several thousand per year.[26] From the fossil record, it is estimated that the normal rate of extinction, when the biosphere is not in crisis, is only about one species every four years. So the biosphere is truly in crisis.

During the Big Five episodes of past extinctions, the biosphere lost 50–80 percent of its marine organisms and 20–70 percent of its terrestrial species. Obviously, modern humans had nothing to do with this since we entered the picture only about two hundred thousand years ago. Climate change, marine estuary habitat loss due to continental coalescence, and one or more asteroid impacts all probably contributed to the Big Five. By contrast, human activities contribute significantly to the present extinction crisis. Preeminent environmental biologist Edward O. Wilson (2006, 118) describes these activities with the acronym HIPPO, where letter order corresponds to the rank of each activity's destructiveness:[27]

H = habitat loss from diverse factors including deforestation and climate change;

I = invasive species that displace native plants and animals;

P = pollution, which has been accumulating since the Industrial Revolution;

P = population (human) excess, which exacerbates the other four causes; and

O = overharvesting of non-domestic food and game plants and/or animals.

If HIPPO activities continue unabated, half of Earth's plant and animal species are projected to disappear by the middle of this century. If humans were to halt all of the above destructive activities within a few decades, estimates based on past major extinctions show that nature's restoration of the lost biodiversity will require about 10 million years. Presuming that humankind survives the sixth extinction, it will be faced with major ecological voids within this century. It is doubtful that humans will patiently wait 10 million years for nature to fill these voids via the evolutionary process. What, if any, role synthetic biology ought to play in species replacement after the sixth extinction is a question that needs careful thought, the sooner the better.

Homo sapiens: *moving from cocreator to creator*. Evolution by natural selection and perhaps other mechanisms was the first and primary creator of living things on Earth over the past four billion years. Humankind has been a cocreator of life forms, along with biological evolution, ever since it began domesticating plants and animals more than ten thousand years ago. Our cocreative activities continue into the present with selective hybridization and breeding, induced mutagenesis, and genetic engineering technologies. But if synthetic biology is successful in generating new organisms from scratch with human-designed genomes, we will have advanced from cocreators to independent creators of life.

Do the human habits of commodifying nature, exempting humans from certain laws of nature, allowing trained incapacities in biology within our education systems, and the ongoing sixth extinction oblige us to work on a special ethic to guide the aims and results of synthetic biology? As previously noted, this is a controversial question, and reasonable persons will disagree on the answer. In my view, prudent vigilance, as urged by the Presidential Commission for the Study of Bioethical Issues (2010) for synthetic biology, is not a sufficient response to synthetic biology. What questions should an ethic for synthetic biology explore? All or a subset of the following five questions could be a starting place.

1. Are there qualitative differences between synthetic biology and genetic engineering?
2. Should a moral boundary be drawn between "natural" and "synthetic" life forms regarding their ethical treatment, such as legislating their protection via a threatened/endangered species act?

3. What, if any, moral principles ought to guide the creation of new life forms?

4. What is life?

5. Should public education, both kindergarten through twelfth grade and higher education, be mandated to prepare young persons to make informed and wise decisions about the future of synthetic biology? If so, how should we accomplish such education?

Developing an ethic to guide humankind-the-creator would be a big project, as is developing a thorough answer to any one of the above questions. In any case, as humankind moves forward with synthetic biology, whether via prudent vigilance, adhering to the precautionary principle, or within the framework of a specially developed ethic, it will require wisdom from diverse sources. Ideas and decision-making collaborations must be sought from biologists, other scientists and engineers, philosophers, bioethicists, business persons, elected lawmakers, lawyers and judges, theologians, and the lay public. Persons from all walks of life have insights into humankind's rightful place in nature and how we ought to participate in the history of life on Earth. In the meantime, we must each do our part to always be becoming better informed about the scientific and ethical issues associated with synthetic biology and all other twenty-first-century biotechnologies.

Conclusion

Synthetic biology, the newest biotechnology, entails collaboration among researchers in several life and physical science disciplines, and its aim is to create new life forms with medical, environmental, military, and industrial significance. Top-down synthetic biology uses existing life forms as a foundational, living scaffold upon which to add genetic traits that serve humankind. It goes far beyond the older genetic engineering technology that adds or alters just one or a few genes; instead, top-down synthetic biology aspires to redesign entire genomes or networks of genes. The result is brand-new life forms with biochemical functions not found in nature's array of living things. Bottom-up synthetic biology aims to create living cells/systems from scratch, perhaps even using genetic codes of entirely human design. In addition to ethical issues raised by other modern biotechnologies, including safety, distributive justice, and the patenting and commercialization of living things, synthetic biology raises new issues surrounding the role of humankind as a creator of life. Prominent among these is the relative value to be placed on life forms and ecosystems created via eons of evolution compared to life forms and ecosystems constructed by humans.

Questions for Thought and Discussion

1. Do you believe that synthetic biology is fundamentally different from genetic engineering? If not, why not? If so, can you think of differences between the two technologies in addition to those described in the text?

2. Do you believe that synthetic biology raises unique ethical problems related to humans as creators of life? If so, what are they? If not, why not?

3. Do you think that synthetic biology research should be regulated by federal or international law? If not, do you think it should be subject to any regulation? If so, what restrictions should be enforced?

4. Should an ethical component be added to the technological component of the iGEM competition? If so, how would you do this?

5. What specific ethical questions do you think emerge from synthetic biology?

6. What role, if any, do you believe public education ought to have in teaching about scientific and ethical aspects of modern biotechnologies?

Notes

1. The ETC Group (Action Group on Erosion, Technology, and Concentration) website is at www.etcgroup.org (accessed May 22, 2012). There it is stated, "ETC Group supports socially responsible development of technologies useful to the poor and marginalized and we address international governance issues affecting the international community. We also monitor the ownership and control of technologies and the consolidation of corporate power." The ETC Group has lengthy, well-researched, and well-written reports on synthetic biology and

nanotechnology available for free download off its website. The definition of synthetic biology comes from the ETC Group (2007) report, "Extreme Genetic Engineering: An Introduction to Synthetic Biology."

2. Parens, Johnston, and Moses (2008) argue that synbio presents no new categories of ethical problems beyond those that arise from other new technologies, i.e., issues of safety, risk, patents, controlling monopolies, and distributive justice.

3. ETC Group's (2007) "Extreme Genetic Engineering: An Introduction to Synthetic Biology" is a sixty-four-page review of synbio and its implications; it contains 281 endnotes and is written in nontechnical language.

4. Boldt and Müller (2008) argue for an extended Asilomar-like dialogue among synthetic biologists to develop a code of ethics for their discipline that reflects an understanding of how synbio activities impact society and nature as a whole.

5. See http://partsregistry.org/Main_Page (accessed May 22, 2012). The home page introduces the registry like this: "The Registry is a collection of ~3500 genetic parts that can be mixed and matched to build synthetic biology devices and systems. Founded in 2003 at MIT, the Registry is part of the Synthetic Biology community's efforts to make biology easier to engineer. It provides a resource of available genetic parts to iGEM teams and academic labs."

6. BioBricks are actually molecular entities: DNA segments like promoters and terminators, genes coding for proteins like repressors and activators of gene expression, and DNA that codes for RNA with regions that tell ribosomes (cellular protein synthesizing machines) how many proteins to make, how quickly to make them, when and where to stop protein synthesis, and how long the life span of a protein will be.

7. The researchers chose *Mycoplasma genitalium* for this study because its genome is the smallest for all organisms that can be cultured in pure form in the laboratory. The expectation is that this set of genes numbers close to that required to sustain life.

8. In this research article (Lartigue et al. 2007), the researchers transplanted the genome of *Mycoplasma mycoides,* a disease agent of goats, into another species of bacterium, *Mycoplasm capricolum.* At first, the host cell contained two genomes,

but after the cells divided, some new cells contained only the transplanted genome. These new cells lived and reproduced perfectly well, assuming the outward appearance of the donor cell. Thus, one species of organism was literally
transformed into another species solely by the swapping of DNA molecules. This
demonstrates the feasibility of creating new life forms simply by putting new, synthetic genomes into cells.

9. The researchers (Smith et al. 2003) synthesized the 5,386-base-pair-long
genome of a virus from scratch using commercially available chemicals. A science news item in *Science* magazine (November 21, 2003, 1307) reports that "Venter is convinced that he can build genomes 300,000 bases or longer." This is still
far short of most bacterial genomes, which typically contain 1–10 million bases.
Gibson et al. 2008 details the use of living yeast cells to aid in the assembly of
human-made fragments of DNA into a complete genome for a bacterial cell. The
next step will be to show that this human-made genome is capable of transforming living cells into organisms with characteristics that reflect the genetic information in the transplanted genome. Then the stage will be set for humans to design wholly artificial genomes for transplantation and the formation of new life
forms.

10. Science reporter Elizabeth Pennisi (2009) explains what Venter and Church
accomplished in relatively nontechnical terms and the significance of the work for
future synthetic biology. The technical report of Venter's work was published in
Lartigue et al. 2009.

11. See http://protocells.lanl.gov/ (accessed October 3, 2011). The Protocell
Assembly project's overarching goal is to discover conditions under which new,
simple life forms will assemble. An international repository for protocell information is at www. protocell.org (accessed October 3, 2011).

12. See http://www.hhmi.org/research/investigators/szostak_bio.html (accessed October 3, 2011). This web page for Jack Szostak describes his work with
protocells and his use of in vitro natural selection to create RNA molecules with
enzymatic activities.

13. Benner, Hutter, and Sismour write that their work is the next step in the
tradition of biomimetic chemistry, a subdiscipline of organic chemistry that attempts to synthesize chemical structures different from natural, biological mole-

cules but that behave similarly to nature's molecules. The goals defining synthetic biology, they write, are "to reproduce advanced, complex, and dynamic behaviors of biological systems, including genetics, inheritance, and evolution" (Benner, Hutter, and Sismour 2003, 125–26).

14. The authors (Yang et al. 2006) describe two new bases similar to nature's A, C, T, and G and report that these have been used to develop diagnostic tools for hepatitis B and C and for detection of genetic defects associated with cystic fibrosis. See also Klotz 2009.

15. This article (Katsnelson 2009) also reports on some of the technical problems encountered by students, which reflect unsolved problems in the synbio discipline as a whole. One major difficulty is that the short DNA sequences at each end of a BioBrick, which serve to "glue" different BioBricks together, often interfere with the operation of the biological devices constructed.

16. A. Campbell 2005 describes and illustrates some of the students' projects in the 2004 iGEM competition at MIT.

17. Parens, Johnston, and Moses (2008) argue that synthetic biology presents no new categories of ethical problems beyond those that arise from other new technologies. For a full report of their study, see Parens et al. 2009.

18. Dyson sees mostly positive things coming from increased accessibility to methods for the genetic manipulation of life: "Domesticated biotechnology, once it gets into the hands of housewives and children, will give us an explosion of diversity of new living creatures, rather than the monoculture crops that the big corporations prefer. New lineages will proliferate to replace those that monoculture farming and deforestation have destroyed. Designing genomes will be a personal thing, a new art form as creative as painting or sculpture" (2007, sec. 1, para. 6).

19. US Patent Application 20030170663 by Claire M. Fraser et al., September 11, 2003. This patent application by Craig Venter and his group for the minimal genome of *Mycoplasm genitalium* is at http://appft.uspto.gov/netacgi/nph-Parser ?Sect1=PTO2&Sect2=HITOFF&u=%2Fnetahtml%2FPTO%2Fsearch-adv.html &r=8&p=1&f=G&l=50&d=PG01&S1=(%22minimal+genome%22+AND+ Venter)&OS="minimal+genome"+and+Venter&RS=("minimal+genome"+AND +Venter) (accessed June 15, 2012).

20. This article (Schmidt et al. 2008) is a compilation of contributions from 124 e-conference participants from twenty-three different countries in six categories: (1) ethics, (2) safety, (3) security, (4) intellectual property rights, (5) governance and regulation, and (6) public perception.

21. Examples include corn plants engineered with bacterial genes to render them resistant to insect pests, beans with daffodil genes to confer resistance to herbicides, rice with bacterial and daffodil genes to add vitamin A to rice's nutritional value, pigs with human cell surface proteins to give pigs potential as organ donors for humans, and bacteria that produce human insulin or growth hormone.

22. An excellent slide presentation explaining the history of recombinant DNA research and the current NIH guidelines is at http://oba.od.nih.gov/oba /IBC/ASGT_2007_Training/Overview%20of%20the%20NIH%20Guidelines .pdf (accessed May 22, 2012).

23. Quoting Wilson (2006), the fifth factor is "invasive species (harmful aliens, including predators, disease organisms, and dominant competitors that displace natives)." It too can be considered a product of nature's commodification in that the interstate and international transport of fresh food products to satisfy humans' appetite for out-of-season produce and to provide profits for the transporters also provides a means for hitchhiking alien species to invade new territories.

24. Synthetic biologist Drew Endy, Department of Bioengineering at Stanford University, is credited with this statement in the ETC Group (2007) report "Extreme Genetic Engineering." In Endy's defense, the web page for his laboratory describes his research in the engineering of genetically encoded memory systems as having potentially profound impacts on the study and treatment of diseases.

25. There are several books with this term in their titles. One of the most accessible of these is by paleoanthropologist Richard Leakey (Leakey and Lewin 1995), *The Sixth Extinction: Patterns of Life and the Future of Humankind.*

26. Estimates of the number of species that will be lost *annually* in this century range from seventeen thousand to one hundred thousand. Estimates differ due to differences in the estimated total number of species on Earth. This rate of extinction, along with the rate at which species enter trajectories toward premature extinction, is about one hundred times the rate at which evolution brings new species into existence. As the remnants of many ecosystems disappear later

in this century, the extinction rate is expected to rise to more than one thousand times the rate at which species are born.

27. The estimated number of species of living things on Earth ranges from 3.6 to 112.0 million. Of these, less than two million are known to biologists.

Sources for Additional Information

Baker, D., G. Church, J. Collins, et al. 2006. "Engineering Life: Building a FAB for Biology." *Scientific American,* June, 44–51.

Chopra, P., and A. Kamma. 2006. "Engineering Life through Synthetic Biology." *Silico Biology* 6: 401–10.

ETC Group. 2007. "Extreme Genetic Engineering: An Introduction to Synthetic Biology." www.etcgroup.org (accessed February 3, 2009).

Leakey, R., and R. Lewin. 1995. *The Sixth Extinction: Patterns of Life and the Future of Humankind.* New York: Random House.

Tucker, J. B., and R. A. Zilinskas. 2006. "The Promise and Perils of Synthetic Biology." *New Atlantis* (Spring): 25–45.

Wilson, E. O. 2006. *The Creation: An Appeal to Save Life on Earth.* New York: W. W. Norton & Company.

Zimmer, C. 2007. "The Meaning of Life." *Seed* (July/August): 68–73.

Conclusion

In an interview for a film titled *The Origin of Species,* from the Great Books series, evolutionary biologist and geneticist George Gaylord Simpson expressed feelings of reservation but also certainty about the eventual eugenic use of human genetic engineering. His words express concern, the importance of knowledge, and inevitability about our future use of biotechnology:

> I am pretty sure that if we survive, if we do not destroy ourselves with pollution, atomic war, and so on, we will sooner or later wish to take control of our own evolution. But I hope we do not do it too soon. I mean, at the moment we are far too ignorant, both of genetics and of what we really want, to tinker with our own evolution. So I'm not urging eugenic measures upon us. I'm really not. I hope we will not do that. But ultimately I'm sure we will. What we shall decide we want, of course, is up to our great-great-great-grandchildren. But they will take this on. (Great Books 1993)

If Simpson is right, the activities of our great-great-great-grandchildren will be consistent with the view of fifteenth-century philosopher Pico della Mirandola, whose words introduced this book and who saw humankind as cocreator of itself. Recalling Pico and his exuberance over human-kind's creative potential, I am moved to return to Michelangelo Buonarroti (1475–1564) and his masterpiece now standing in the Galleria dell' Academia in Florence, Italy (see fig. 6.1). In his PBS film on Florence and the Renaissance, Bill Moyers had this to say about the *David* and the legacy of fifteenth-century humankind's decision to grasp control of its own fate:

> The heroic ideal of David was the ultimate expression of Florentine optimism—their belief in the divinity of the human being. Created in the image of God, now Man was creating images of himself. Contained in the marble was the mirror of the soul. We needed only to free it as Michelangelo had liberated David from the block. But what does it bring, this gift of freedom? Certainly the ability to dream and the power to achieve. But perhaps, as Michelangelo had seen, it also leaves us standing alone—to make and face our fate. (PBS Home Video 2008)

As did Michelangelo, we twenty-first-century voyagers through time also have the freedom to create new images of ourselves. The nature of our dreams and how we apply our power to achieve will shape our own futures and the future of our species. And as the most creative minds of the Renaissance sensed, we stand alone to shape our fate and then face what we have formed. Accepting the responsibility of informed decision making is the foundational step we must all make into humanity's future.

Science writer Joel Garreau (2005) interviewed dozens of persons directly associated with the development of genetic, robotic, informational, and other technologies that together have the potential to transform humanity for better or worse. He uncovered three views about the future of humanity: the heaven scenario, the hell scenario, and the prevail scenario. The heaven scenario is that these technologies will usher in a new era of well-being for humanity, where disease, mortality, poverty, and war are thrown into the dust heap of history. The hell scenario is that we will lose the essence of our humanity to these technologies, and humanity as we know it now will come to an end. In the prevail scenario, humanity's future is a collage of mistakes and successes. But in the end, according to the prevail scenario, we will use twenty-first-century biotechnologies for the overall well-being of the species. My hope is that each of us will work with an informed mind toward that end.

Epilogue

Cell and molecular biology, genomics, neuroscience, and the biotechnologies emerging from these basic sciences advance at breakneck pace. Minutes after sitting down to write this section about keeping up with such rapidly developing technologies and what we can do about it, my wife walked into the room and asked, "Did you hear on NPR this morning about the new drug trial for progeria?" "No I didn't," I said, "because I've been busy here trying to update my book on biotechnology." "Well, you need to know about it," she said, "because it may lead to a way to retard aging in people like you and me." What timing! This is just the problem. Even as I work to update myself in one area of biotechnology, something big is happening in another area. I spend time every day trying to keep up; but what about you students, young parents, nonscientist professionals, and other busy citizens whose days already brim with essential activities claiming time and attention? How can you stay informed about developments in twenty-first-century biotechnologies?

There is some bad news, but mostly good news. The bad news is that virtually nobody can stay absolutely current when it comes to biotechnology. Reading a book like this provides a foundation for keeping informed. But scientific details in any newly published book about biotechnology are bound to be three or more years out of date, compared to scientists' current research, since the route between scientific discoveries at laboratory benches and books is long and winding.

Now for the good news. There are diverse and handy sources for current news about biotechnology and its implications for individuals and society. Here are some:

American Association for the Advancement of Science (AAAS): See http://www.aaas.org/news/ (accessed May 16, 2012). This AAAS page contains links to current news about scientific discoveries that affect us all, including many on biotechnology and its associated ethical and societal issues.

Ethical, Legal, and Social Implications (ELSI) of the Human Genome Project (HGP): An extensive suite of pages and links to articles and other websites with current information on the ethical implications of diverse biotechnologies is at this section of the government's HGP site: http://www.ornl.gov/sci/techresources/Human_Genome/elsi/elsi .shtml (accessed May 16, 2012).

Genetics and Public Policy Center, Johns Hopkins University: Current and archived by year, news releases on ethical and legislative aspects of genetic biotechnologies, including assisted reproduction, prenatal genetic testing, genetic therapy trials, stem cells, direct-to-consumer genetic testing, and personalized medicine, are at http://www.dnapolicy.org /news.release.php?year=2011 (accessed May 16, 2012).

National Human Genome Research Institute: Free information, including transcripts of lectures, about the human genome project and its applications are accessible from the HGP home page at http://www.genome .gov/About/ (accessed May 16, 2012).

National Public Radio: *Science Friday,* a weekly, afternoon call-in talk show with scientists, hosted by science journalist Ira Flatow and underwritten by the National Science Foundation. Each broadcast consists of two one-hour programs about nature, science, and technology. Past programs are archived as podcasts at http://www.sciencefriday.com/ (accessed May 16, 2012).

National Science Foundation (NSF): Links to short, nontechnical descriptions of NSF-funded research in all scientific disciplines are on the "Discoveries" page of NFS's website at http://www.nsf.gov/discoveries/ (accessed May 16, 2012).

NOVA/PBS: NOVA programs aired on public television stations revolve on the premise that the world of science is fun. Programs on special topics air regularly on local PBS stations. Program transcripts and audio and video recordings of material related to program topics are freely available at http://www.pbs.org/wgbh/nova/ (accessed May 16, 2012). From this site one can hear experts discuss scientific and ethical aspects of cloning, stem cell research, genetic engineering, age retardation, neuroscience, DNA and human evolution and diversity, human genome projects, gene therapy, nanotechnology, and synthetic biology.

Periodicals, including *Discover Magazine, Scientist: Magazine of the Life Sciences, Science,* and *Nature,* publish nontechnical articles and news items about current developments in biotechnology and its ethical aspects. These all have their own websites with links to free articles, news releases, podcasts, and blogs. Hard copies are also at most libraries and many bookstores.

The problem of datedness in this book is blunted in several ways. I incorporated new research results reported in the scientific literature and political events especially relevant to the book's topics regularly during the years of its writing.

When deciding what scientific information to include in the book, I gave priority to foundational science that will remain crucial for understanding future developments in biotechnology. For example, new methods for obtaining *pluripotent* cells may appear over the years, but basic biological properties of pluripotent cells will not change. Their potential to give rise to the body's diverse tissue types will remain, so I emphasized that concept. Finally, discussions about human values associated with each biotechnology are timeless. A case in point is personhood, specifically, the problem of defining the moral status of a human embryo. This issue will remain with us regardless of future advances in stem cell technologies, and reasonable persons will continue to disagree. Since appreciating various positions on personhood is important for productive public dialogue and wise policy formation, I devoted some pages to this subject. The same applies to arguments for and against cloning, genetic enhancement, neuroenhancement, age retardation, embryo selection, and forging ahead with synthetic biology.

I conclude with a small sampling of noteworthy developments in biotechnology that occurred too late to include in the text:

Age retardation: This is heralded by *Science* as a runner-up for the 2011 scientific Breakthrough of the Year. Researchers at the Mayo Clinic in Minnesota and in the Netherlands used a drug to selectively kill senescent cells from the body of mice. Drug treatment delayed several maladies of aging, including loss of muscle and fat tissue and decreased fitness. Future research will explore the drug's potential to retard aging in humans. See Peeper 2011; Baker et al. 2011; and Breakthrough of the Year 2011.

Prenatal diagnosis: In 2012 researchers led by Jay Shendure at the University of Washington in Seattle reported the noninvasive whole-genome sequencing of a fetus from a sample of the mother's blood (Kitzman et al. 2012). The sample was drawn 18.5 weeks into the woman's pregnancy, and from it the researchers identified and sequenced fetal DNA circulating in the mother's blood. Over three thousand disease-causing mutations are able to be detected by the procedure, which eliminates the risks of amniocentesis and chorionic villus sampling.

Stem cells: In just five weeks, workers at Stanford University converted human skin cells directly into functional nerve cells, skipping the step of first transforming skin cells into *iPS cells*. The work could ultimately give rise to therapies for patients with Parkinson's disease and other nerve cell pathologies, and it avoids the personhood issue that plagues ESC research (Conger 2011). And in 2012 two laboratories reported remarkable success using iPS cells to repair heart tissue in live mice that was previously injured by myocardial infarction (heart attack). The researchers injected viral vectors carrying three or four genes for transcription factors that regulate heart development directly into the injured area of the heart. The vectors carried the genes into cardiac fibroblasts (non-muscle cells), which then transformed the fibroblasts into beating, cardiac muscle cells, resulting in an overall improvement of heart function (Palpant and Murry 2012).

Gene therapy: In 2011 researchers at the University of Pennsylvania Abramson Cancer Center and Perelman School of Medicine achieved a breakthrough in using gene therapy against cancer. They removed immune system cells from patients with chronic lymphocytic leukemia (CLL), genetically engineered them to seek out and destroy the cancer cells, and then reintroduced the engineered cells into the patients. The workers also engineered the cells to proliferate profusely after attaching to tumor cells. In a pilot trial, more than two pounds of tumor were destroyed within three weeks in patients injected with the engineered cells, and the patients remained cancer free for the several months prior to publication of the trial's results. Long-term safety of the procedure must still be established, but the researchers are optimistic that this type of gene therapy will lead to effective treatments for CLL and also for ovarian, pancreatic, lung, myeloma, and melanoma cancers (Paddock 2011).

In 2012 a UK ethics council approved a form of germ-line gene therapy, transferring the nucleus from an IVF patient's egg into the enucleated egg from another woman in order to cure a disease carried by the mitochondrial DNA of the patient's egg (Zielinska 2012). The resulting embryo, and infant, would carry genetic material from two different women—nuclear DNA from the patient and mitochondrial DNA from the second woman. Recall that mitochondrial DNA is inherited only through eggs, not through sperm. Although the procedure has not yet been used, the action by the ethics council paves the way for legalizing the technique.

Synthetic biology: In 2011, Cambridge University researchers genetically modified a nematode worm so that it can make *proteins* containing an unnatural *amino acid*. The work involved altering the genetic code of the animal so that one of its three-letter *codons* that normally does not code for any amino acid now codes for the unnatural amino acid. The procedure, called genetic code expansion, was developed in bacteria several years earlier, but this marks the first use of the technology in a multi-celled animal. Theoretically, the procedure could be applied to any organism, and it makes it possible for living cells to produce proteins with properties not found in nature (Greiss and Chin 2011).

Pico della Mirandola ([1486] 1956, 287) speaks with relevance through six centuries: "Thou, constrained by no limits, in accordance with thine own free will . . . shalt ordain for thyself the limits of thy nature . . . with freedom of choice and with honor . . . thou mayest fashion thyself in whatever shape thou shalt prefer."

Glossary

Note: The glossary contains definitions for italicized terms in the text. Italicized terms within definitions are also glossary entries.

ADHD: Attention deficit/hyperactivity disorder, trouble with inattentiveness, over activity, impulsivity, or a combination of these.

adult stem cell: A category of *stem cell* found in tissues and organs and whose normal function is to replace dead cells within that tissue or organ and which replenish themselves by *mitosis.*

age retardation: Slowing down the rate of accumulation of undesirable changes to the mind and body that accompany the passage of time.

amino acid: Chemical building blocks that are joined together into a chain to produce a protein; twenty different amino acids are used by the *cell* to make proteins.

asexual reproduction: Procreation without involvement of germ cells (eggs and sperm); typically, a part of the parent organism gives rise to an entirely new organism genetically identical to the parent, as when a cutting from one plant develops into another new plant.

ATP: Adenosine triphosphate, the energy currency of the cell; this energy-rich molecule can be oxidized to provide the energy needed for virtually all of the activities of a living cell, including movement, reproduction, and synthesis of other types of molecules.

axon: A long extension of a nerve cell that transmits electrical signals down its length to other nerve cells.

base: A chemical subunit of *DNA* and *RNA*; designated A, T, G, and C in DNA and designated A, U, G, and C in RNA.

base complementarity: The pairing of specific *bases* opposite each other and present in different strands of a double-stranded DNA or RNA molecule; members of each base pair (A and T, A and U, C and G) form weak chemical bonds with each other that help hold the two strands of the double-stranded DNA or RNA molecule together.

base sequence: Refers to the linear order of chemical subunits (*bases*), for example, A, G, C, and T or U, in a *DNA* or *RNA molecule*.

biodiversity: The array of species of living organisms present in a specified area.

blastocyst: An early-stage mammalian embryo that may implant itself into the wall of the uterus, where it can give rise to the fetus and the fetal components of the placenta.

blastomere: A single *cell* in a cleavage-stage embryo; in mammals these cells comprise embryos containing two to sixteen cells.

cell: The fundamental structural and functional unit of all living things on Earth.

cell differentiation: Cellular process(es) during development by which cells become specialized for particular tasks in specific tissues and organs, for example, liver, heart, skin, and nerve.

central dogma of biology: The concept that genetic information stored in DNA is used to produce a messenger RNA molecule, and the information in the RNA is then used to produce protein; it describes the directional flow of genetic information in cells: from DNA to RNA to protein.

chemical synapse: A juncture between two neurons at which *neurotransmitter* molecules are released.

chimera: An organism that from early development is comprised of cells originally present in two or more organisms; it differs from a genetic hybrid in that any given cell in a chimera, and the entire set of genes in that cell, is from just one particular organism, whereas the gene sets in all cells of a genetic hybrid are a mixture of genes from two different organisms.

chromosome: A single, double-stranded molecule of *DNA* and various adhering *proteins* in either a *prokaryotic* or *eukaryotic* cell.

cilium/cilia: Filamentous projections of the cell membrane with an internal structure comprised of *microtubules* that produce whip-like motion, resulting in *cell* movement and/or directional movement of material along the outer surface of cells; a sperm tail is a specialized cilium called a flagellum.

cloning: A generic term for several different procedures for diverse purposes: making copies of *DNA* fragments of interest (DNA cloning, molecular cloning, or gene cloning), producing embryos genetically identical to another embryo or organism (*therapeutic cloning* or *embryo cloning*), and creating entire organisms (plants or animals) genetically identical to a particular, preexisting embryo or organism (*reproductive cloning*).

codon: A sequence of three contiguous *bases* (A, U, G, C) in a *messenger RNA* that specifies a particular *amino acid* in the *protein* coded for by that messenger RNA; for example, the AAG codon specifies the amino acid lysine.

conception: Earliest stage in human development, marked by penetration of an egg cell by a sperm cell.

cytoplasm: The material of a *cell* enclosed by its outer cell *membrane,* excluding the *nucleus*; within the cytoplasm are all of the cell's *organelles* except the nucleus and the few organelles inside the nucleus.

cytoskeleton: The array of *microtubules* and various types of filaments, including *microfilaments,* inside *eukaryotic cells* that function to maintain cell shape and in-cell motility.

dendrite: A thin extension of a *neuron* that receives incoming signals from other neurons.

de novo: Latin expression meaning "from the beginning" or "brand new."

developmental potential: Usually applied to single *cells,* but could also apply to groups of cells; refers to the array of different cell types to which a cell or cells are capable of giving rise, for example, the developmental potential of the fertilized egg is complete in that it can give rise to all cell types in the body.

diploid: A genetic state characterized by *cells* having two complete sets of *chromosomes,* normally one paternal set and one maternal set; for example, body cells, excluding eggs and sperm, are diploid.

distributive justice: The state or process of resources or benefits being made available to people in a fair way.

DNA: 2-deoxyribonucleic acid, the double helix *molecule* that contains the genetic information for an organism; a single molecule of double-stranded DNA comprises each *chromosome.*

DNA sequence: Refers to the *base sequence* or linear order of bases (A, G, T, C) in each strand of a double-stranded DNA molecule; when the base sequence for one strand of a double-stranded DNA molecule is known, the base sequence of the other strand can be inferred by the rules of *base complementarity.*

DNA sequencing: Laboratory procedures for determining the linear order of *bases* (A, G, T, C) in *DNA molecules.*

ectoderm: The outermost of the three germ layers in the gastrula-state embryo; its *cells* are committed to forming structures like the skin, lens of the eye, and epithelial cell lining of the nose, mouth, and anal canal.

elective abortion: The induced termination of a pregnancy, usually before the fetus is viable outside the uterus (before the beginning of the third trimester), at the decision of the woman carrying the fetus and performed at her request.

embryo cloning: See *therapeutic cloning.*

embryonic germ cell: A *pluripotent* cell derived from *primordial germ cells* in the *gonadal ridge* of four- to seven-week-old human *embryos* or *fetuses.*

embryonic stem cell: A *pluripotent* cell derived from the inner cell mass of a *blastocyst*-stage embryo.

embryonic stem cell line: A population of *embryonic stem cells* derived from one *blastocyst.*

endoderm: The innermost of the three germ-layers of the gastrula-stage embryo; its cells are committed to forming the liver, pancreas, and the inner lining of the gut and respiratory passages.

enzyme: A class of *proteins* that aid cells in carrying out chemical reactions necessary for life.

epigenetic: Refers to a trait or phenomenon caused by or inherited independently of the actual *DNA base sequences* in an individual's *chromosomes*; can refer to traits controlled by the *epigenome*.

epigenome: A pattern of chemical modifications to certain *DNA bases* and certain *proteins* associated with *eukaryotic DNA;* the modifications influence the activity of *genes*.

ESC: *Embryonic stem cell.*

ESC line: *Embryonic stem cell line.*

eugenics: The science of improving the human species by the careful selection of parents and/or by discouraging or preventing mating between certain parental types.

eukaryote: Any organism belonging to the kingdom Eukarya; organisms comprised of *eukaryotic cells.*

eukaryotic cells: *Cells* that contain a *nucleus* and other *membrane bound organelles*; all plants and animals readily visible to the unaided eye as well as many microscopic organisms are comprised of one or more eukaryotic cells; these are thought to have arisen from *endosymbiotic* relationships between preexisting *prokaryotic cells* about three billion years ago.

fertilization: Stage in development marked by fusion of the paternal and maternal nuclei from the egg and sperm, respectively, also called *syngamy*.

functional genomics: Analysis of the *base sequences* in the *genome* of an organism to determine the function of specific genes.

gene: Generally refers to a region of a *DNA molecule* that contains information to make an *RNA* molecule; RNAs from some genes code for specific *proteins*; some *genes* do not specify RNA molecule formation but instead regulate the activity of other genes.

gene expression: The process whereby the information in *DNA* is used to produce *mRNA* and the information in the mRNA is used to produce a *protein*; when a protein is being produced inside a cell, we say that the *gene* for that protein is being expressed.

genetically engineered organisms: Plants, animals, or bacteria that have had their *genomes* purposefully altered using laboratory procedures that add, delete, or modify *genes*; often genes from two or more different species are combined in one genome.

genetic code: The enciphering system in *DNA* and *RNA* that specifies the synthesis of specific proteins.

genetic engineering: Human modification of an organism's *genome* using twentieth- and twenty-first-century methods of *DNA* technology; often involves introducing one or a few foreign or synthetic *genes* into an organism's genome, as distinguished from the wholesale overhaul of genomes via *synthetic biology*.

genetic enhancement: Use of *genetic engineering* technology to augment genetically specified traits.

genetic recombination: Exchange of corresponding segments of *DNA* in maternal and paternal *chromosomes* during *meiosis,* resulting in new combinations of *gene* variants in eggs and sperm; also called chromosomal crossover or crossing over.

genome: The total *DNA* in a cell's *chromosomes.*

genomics: Identification and genetic analysis of the *base sequence* in all of the *DNA* for an organism; includes learning the detailed chemical structure of *genes* but not necessarily the function of genes.

genotype: The genetic constitution of an individual, either in terms of a single trait, a well-defined group of traits, or the entire *genome.*

germ cell: An egg or sperm cell; responsible for producing new generations of sexual organisms.

gonadal ridge: A region of tissue in five- to ten-week-old human embryos or fetuses that contains *primordial germ cells,* cells that will give rise to eggs or sperm in the fully formed individual.

haploblock: A locus (location) in the *genome* containing several thousand *bases* and twenty to seventy *SNPs (single nucleotide polymorphisms)* that are all inherited together from one generation to the next; during *meiosis,* the exchange of genetic material via chromosomal crossing over occurs in units of haploblocks.

haploid: A genetic state characterized by *cells* that have only a single set of *chromosomes*; for example, sperm and eggs cells are haploid.

haplotype: A variant of a particular *haploblock* characterized by a unique pattern of *SNPs*; there are generally fewer than ten haplotypes for each haploblock in humans.

HapMap: "A catalog of common genetic variants that occur in human beings. It describes what these variants are, where they occur in our *DNA,* and how they are distributed among people within populations and among populations in different parts of the world." (Definition from the official International HapMap Project website: http://hapmap.ncbi.nlm.nih.gov /whatishapmap.html [accessed April 29, 2012]).

HGP: *Human Genome Project.*

high speed centrifugation: A laboratory technique for separating and isolating *cell organelles* and other subcellular components for further study; cells are ruptured, their contents are suspended in a liquid medium inside a glass or plastic tube, and the tube is then spun at a nearly horizontal angle at high g-forces; the various cell parts sediment sequentially at the bottom of the tube depending upon their size and density.

Human Genome Project (HGP): The international, publicly funded project to fully decipher the arrangement of subunits that comprise human *DNA,* thereby learning the structure of all of the genetic information necessary for development of a human being.

individuation: Refers to an embryo developing to the stage where it can no longer split to give rise to identical twins; in humans this happens at about fourteen days after fertilization.

induced pluripotent stem cells: *iPS cells; somatic cells* made *pluripotent* by the introduction of one or more genes into their genomes; the first human iPS *cells* were derived from skin fibroblasts in late 2007, and in 2009 human iPS cells were derived from neural stems cells; iPS cells are an alternative to *embryonic stem cells* for development of clinical applications. RNA-induced plutipotent stem cells are another type of iPS cell created in the laboratory by introducing certain mRNA molecules into the cytoplasm of somatic cells.

inner cell mass: A group of *pluripotent cells* inside four- to seven-day-old human *blastocysts* from which *embryonic stem cells* are derived and which can give rise to a fetus once the blastocyst is implanted in the uterine wall.

in utero: Inside the uterus.

in vitro: Describes any laboratory procedure where living cells are grown or maintained outside the environment of a living organism; it is Latin for "in glass."

in vitro fertilization (IVF): A laboratory procedure for uniting eggs and sperm outside of the body to produce embryos that can be implanted into the uterus for development to term or frozen for later implantation or research; the term *test-tube baby* refers to the baby's conception by IVF.

iPS cells: *Induced pluripotent stem cells.*

IVF: *In vitro fertilization.*

liberal eugenics: An ideology that encourages, or at least permits, using reproductive and/or genetic technologies such as *in vitro fertilization, preimplantation genetic diagnosis,* embryo selection, and *genetic engineering* to ehance human traits and capacities and where the choice of such enhancement is left up to individual parents as consumers rather than being promoted by governments or other institutions.

life expectancy: The amount of time a person of a particular age is likely to remain living, based on calculations from mortality statistics for the general population and taking into account occupation and health habits.

life extension: Lengthening vigorous, healthy *life spans.*

life span: The length of lifetime for an individual.

ligand: A generic term for a *molecule* that binds to a *receptor* protein; usually, receptor-ligand binding elicits some physiological response.

lipid: A biological molecule important for forming cellular membranes; examples include saturated and unsaturated fats and cholesterol.

lysosome: A *membrane*-bound vesicle inside the cell that contains digestive *enzymes.*

major depressive disorder: Severe and persistent condition associated with emotional (e.g., sadness, hopelessness, irritability) and physical (e.g., sleeplessness, inability to concentrate, fatigue) symptoms but which is treatable with antidepressant drugs.

meiosis: The process whereby a *diploid cell* undergoes *chromosome* replication and two cell divisions to give rise to four *haploid* cells; sperm and egg cells are produced by meiosis.

membrane: A very thin, semipermeable layer of *lipid molecules* that encloses the contents of the cell as a *plasma membrane* and also encloses various small compartments inside the cell; the types of material that can pass across a membrane are highly regulated by carrier molecules embedded in the membrane and also by the permeability of the membrane itself.

membrane-bound organelle: An *organelle* that contains one or more *membranes* that help to compartmentalize biochemical functions.

mesoderm: The middle of the three germ layers in gastrula-stage embryos; its cells are committed to forming skeletal and heart muscle and the circulatory system.

messenger RNA: mRNA, an *RNA molecule* carrying the information required for the synthesis of a specific *protein.*

microfilament: A type of *nonmembrane-bound organelle* contributing to the *cytoskeleton* of *eukaryotic cell*s; very thin filaments that form bundles and mesh-works responsible for shape changes in the *plasma membrane* that cause cell movement or contraction.

microtubule: A type of *nonmembrane-bound organelle* contributing to the cytoskeleton of eukaryotic *cells*; minute, hollow tubes of indefinite length involved in the movement of certain membrane organelles within the cytoplasm and with the separation of duplicated *chromosomes* during *mitosis,* prior to cell division.

mitochondrion/mitochondria: A *membrane-bound organelle* that oxidizes food molecules and converts the energy that is released into ATP (adenosine triphosphate), a form of chemical energy used by the cell.

mitosis: The process whereby a *cell*'s replicated *chromosomes* separate, partitioning themselves equally between two daughter cells produced by cell division; for example, a fertilized egg gives rise to an individual via many mitotic cell divisions.

molecule: An assembly of *atoms* with specific chemical properties different from any of its individual, constituent atoms; molecules combine or associate with each other to form larger molecules or other structures such as *cell membranes, chromosomes,* bone, cartilage, muscle, teeth, and hair.

morula: An early-stage mammalian embryo comprised of about thirty-two unspecialized *cells* packed together into a solid ball of cells; if a morula is split into two pieces, each piece can develop into a whole organism.

mRNA: *Messenger RNA.*

multipotent: Describes *stem cells* in animal organs that can give rise to more than one, but not all, cell types in the body.

nanotechnology: Human manipulation of matter at the level of 0.1 to 100 nanometers, one nanometer equaling one billionth of a meter.

natural selection: The preferential reproduction of individuals with traits best suited to keeping them healthy and alive long enough to successfully reproduce.

nerve impulse: An electrical signal passed down the length of an axon via a flow of sodium and potassium ions across the *cell membrane,* also called an action potential.

neurotransmitter: A small *molecule* released into the *synaptic cleft* by a *presynaptic cell* that interacts with a *postsynaptic cell* membrane to stimulate or inhibit the propagation of a *nerve impulse*; there are many types of neurotransmitters, each with particular functions in the central nervous system.

nonmembrane-bound organelle: An *organelle* that does not contain a *membrane*; for example, *ribosomes, microtubules, microfilaments.*

nucleus: The membrane-bound compartment of *eukaryotic cells* containing the genetic material (*DNA*).

organ: A functionally and morphologically identifiable unit of the body of an organism that is comprised of specific types of *tissues.*

organelle: Literally, "tiny organ"; a small, *cytoplasmic* component of a cell that is specialized for a particular function; there are two types of organelles: *membrane-bound organelles* and *nonmembrane-bound organelles.*

oxygen free radical: Highly reactive *molecules* produced as a normal byproduct of metabolism and which contain unpaired electrons associated with oxygen atoms; since electrons "like" to be in pairs, unpaired electrons "look" for other molecules from which to grab electrons; reactions of oxygen radicals with other molecules may be partly responsible for certain degenerative diseases, cancers, and aging.

personal genomics: Analysis of the *base sequence* of an individual's *DNA* to assess disease risk, drug efficacies and probable side effects, and other medical, physical, or behavioral predispositions.

personalized medicine: Medicine tailored to the specific individual, based on the individual's genetic constitution as it relates to drug responses.

PGD: *Preimplantation genetic diagnosis.*

photosynthesis: The biochemical process whereby green plants use carbon dioxide, water, and sunlight to produce carbohydrates.

plasma membrane: The thin, semipermeable, outer covering of a cell that separates the internal contents of a cell from the external environment yet allows for exchange of material and passage of signals between the cell and the external environment.

pluripotent: Describes animal *stem cells* that can give rise to all of the cell types found in the adult body; only cell types comprising the embryo's contribution to placental tissue cannot be derived from pluripotent cells.

postsynaptic cell: A nerve cell receiving electrical/chemical information from a *presynaptic cell* at the site of a *synapse.*

precautionary principle: An approach to decision making on a potential action that may adversely affect human or environmental health but for which solid scientific evidence about the nature or extent of the action's effects is still lacking; the several formulations of the principle place responsibility for demonstrating the action's safety on those proposing the action, and it urges restraint when knowledge about consequences of the action is lacking.

preimplantation genetic diagnosis: Using *preimplantation biopsy* to screen embryos generated by *in vitro fertilization* for genetic diseases, allowing assisted reproduction technicians to transfer only healthy embryos to a woman's uterus.

presynaptic cell: A nerve cell sending electrical/chemical information to a *postsynaptic cell* at the site of a *synapse.*

primitive streak stage: In humans, a stage of development occurring about fourteen days after *fertilization,* when *cells* that will form the nervous system and other organ systems are first identifiable; the latest stage at which the embryo can spontaneously split into two parts to give rise to identical twins.

progenitor cell: A *cell* derived from a *stem cell* and that is committed to give rise to a particular type of specialized cell.

prokaryote: *Prokaryotic cell.*

prokaryotic cells: Bacteria and blue-green algae; they possess no *nucleus* or *membrane-bound organelles*; earliest known type of cell to exist on Earth about 3.8 billion years ago.

protein: A polymer of *amino acids* joined together in a linear sequence that is determined by the chemical structure of its corresponding *gene.*

proteomic research: Proteomics; the study of *protein* structure and function, including how proteins interact with each other and with other kinds of *molecules.*

psychoactive drugs: Chemical agents that act on the central nervous system to alter brain function and affect cognitive abilities, perception, mood, behavior, and/or consciousness; these may be used therapeutically to treat mental disease, recreationally, or as enhancers of normal brain function.

receptor protein: A *protein molecule* embedded in a *cell's plasma membrane* and which elicits particular cellular biochemical responses when it binds to other specific molecules (*ligands*) in the cell's environment; receptor proteins are communication links between the outside and inside of the cell.

recessive genetic disorder: An inherited disease or other abnormal condition whose expression requires the presence of two copies of a mutated or dysfunctional *gene,* one from each parent.

regenerative medicine: The application of *stem cell* technologies to repair or replace damaged or diseased *tissues* and *organs*.

reproductive cloning: Technology aimed at creating a fully developed individual who is virtually genetically identical to an existing individual.

ribosome: A *nonmembrane-bound organelle* responsible for *translation* of the genetic information in *messenger RNAs* to produce *proteins* in both *prokaryotic* and *eukaryotic cells*; located in the *cytoplasm*.

RNA: Ribonucleic acid; a molecule chemically related to *DNA*; one type of RNA, *messenger RNA,* carries the genetic information in DNA to the site of *protein* synthesis in the *cytoplasm*; other types of RNA, including transfer RNA and ribosomal RNA, function directly in protein synthesis.

RNAi: *RNA interference.*

RNA interference: Inactivation of a specific *messenger RNA* (mRNA) by *RNA* molecules that are complementary to the target mRNA; the latter may be introduced into cells by researchers and provide a new approach to treating cancer and viral infections including HIV and hepatitis.

savior sibling: A child conceived and born, at least in part, to save the life of a sibling needing immunologically compatible blood or other *tissue* to treat a genetic disease.

SCNT: *Somatic cell nuclear transfer.*

selective serotonin re-uptake inhibitor: A class of anti-depressant drugs, including fluoxetine (trade name Prozac), used to treat depression, obsessive-compulsive disorder, panic attacks, and some eating disorders; also called *SSRIs,* they act to block the removal of the neurotransmitter serotonin from *synaptic clefts* and may also stimulate the formation of new *synapses* in the brain.

sexual reproduction: Involves an exchange of genetic material (*DNA*) between different individuals resulting in recombination of *genes* into new mixtures at each generation.

single nucleotide polymorphism: A major source of genomic variation between individual human beings, sites in the *genome DNA* where the same base does not appear in all individuals but varies.

SNPs: *Single nucleotide polymorphisms.*

somatic cell: Any body *cell,* excluding eggs and sperm.

somatic cell nuclear transfer: A method for *cloning* animals; a *nucleus* from a body cell is introduced into the *cytoplasm* of an egg whose own nucleus has been removed, then the egg is activated chemically or electrically and proceeds to develop.

SSRI: *Selective serotonin re-uptake inhibitor.*

stem cell: A relatively undifferentiated *cell* that can give rise to one or more specialized types of cells by multiplying via mitotic cell divisions; the most versatile stem cells are *embryonic stem cells,* which can give rise to every cell type in an adult organism.

surplus embryo: Embryos generated by *in vitro* fertilization that are not transferred to a woman's uterus; they may be stored frozen, discarded, or used for research.

symbiont: An organism living inside or in close association with another organism; each gives and receives benefits from the other organism.

symbiosis: An intimate living arrangement between two organisms; if one benefits at the expense of the other, the symbiosis is parasitic, and if both benefit, the symbiosis is mutualistic.

synapse: A site of communication between a *presynaptic cell* and a *postsynaptic cell* where the *axon* terminal of the presynaptic cell comes very close to a *dendrite* of a postsynaptic cell; in a *chemical synapse, neurotransmitters* are released by the presynaptic cell into the space between the two cells.

synaptic vesicles: Small, *membrane*-bound spherules containing *neurotransmitter* molecules.

synbio: *Synthetic biology.*

syngamy: *Fertilization*; fusing of the egg and sperm nuclei inside an egg *cell.*

synthetic biology: An interdisciplinary endeavor to create new life forms to improve the lives of humans; engineers, biologists, chemists, and physicists all contribute to the endeavor.

telomere: Structure at the tip of *eukaryotic cell* chromosomes that protects internal regions of the *chromosome* from damage but which shortens slightly every time the cell's *DNA* replicates prior to cell division; telomere shortening is associated with aging and cell death.

TEM: *Transmission electron microscopy.*

therapeutic cloning: Technology aimed at producing an embryo in the laboratory that is genetically identical to a patient needing *embryonic stem cell (ESC)* therapy; the embryo would be used to derive ESCs that are fully compatible as transplants for the patient.

tissue: An organized assemblage of several types of *cells,* often a structural and functional unit of an *organ*; for example, muscle, nerve, connective, epithelial, blood, endocrine.

totipotent: Describes *cells* that can give rise to all cell types of an organism; normally, only eggs and sperm, united to form the fertilized egg, are totipotent.

transcription: The process by which genetic information in *DNA* is transferred to an *RNA* molecule.

translation: The process by which genetic information in an *RNA molecule* is used to make a *protein.*

transmission electron microscopy: Method that cell biologists use to visualize cellular structures too small to be seen by conventional light microscopy; a thin section of biological material is stained with lead, uranium, and tungsten, which bind to proteins and lipids, then the specimen is examined with an electron beam focused by magnets; electrons pass through the specimen wherever heavy metals are not bound and produce an image on a screen beneath the specimen.

vectors: Carriers of new *genes* being introduced into *cells* for therapeutic or experimental purposes; disabled viruses are commonly used vectors.

zygote: The fertilized egg *cell* that by serial cell divisions will give rise to every cell in the organism; normally, the zygote is *diploid,* having one paternal and one maternal set of *chromosomes* derived, respectively, from the sperm and egg.

References

Aaroe, J. T. Lindahl, V. Dumeaux, et al. 2010. "Gene Expression Profiling of Peripheral Blood Cells for Early Detection of Breast Cancer." *Breast Cancer Research* 12 (1): R7.

Agency for Healthcare Research and Quality, US Department of Health and Human Services. 2000. "Addressing Racial and Ethnic Disparities in Health Care: Fact Sheet." http://www.ahrq.gov/research/disparit.htm (accessed September 5, 2009).

———. 2008. "National Healthcare Disparities Report." http://www.ahrq.gov/QUAL /nhdr08/nhdr08.pdf (accessed September 6, 2009).

Aiuti, A., F. Cattaneo, S. Galimberti, et al. 2009. "Gene Therapy for Immunodeficiency Due to Adenosine Deaminase Deficiency." *New England Journal of Medicine* 360 (5): 447–58. American Association for the Advancement of Science. 2011. "Evolution on the Front Line." http://www.aaas.org/news/press_room /evolution/ (accessed March 26, 2011).

American-Israeli Cooperative Enterprise. 2012. "Ashkenazi Jewish Genetic Diseases." Jewish Virtual Library, Victor Center for Jewish Genetic Diseases, Philadelphia, PA. http://www.jewishvirtuallibrary.org/jsource/Health/genetics.html (accessed September 8, 2009).

American Medical Association (AMA). 1994. CEJA Report D-I-92, "Prenatal Genetic Screening." http://www.ama-assn.org/resources/doc/code-medical-ethics/212a .pdf (accessed June 7, 2012).

———. 2001. "Principles of Medical Ethics." http://www.ama-assn.org/ama/pub /physician-resources/medical-ethics/code-medical-ethics/principles-medical -ethics.page (accessed June 30, 2012).

American Psychiatric Association. 2000. *Diagnostic and Statistical Manual of Mental Disorders.* 4th ed. Text Revision (DSM-IV-TR). PsychiatryOnline.com. http://dsm.psychiatryonline.org/book.aspx?bookid=22 (accessed October 22, 2009).

Ansari, A. 2010. "Neanderthal Genome Shines Light on Human Evolution." CNN, May 7. http://www.cnn.com/2010/TECH/science/05/07/neanderthal.human.gengen/index.html (accessed July 30, 2010).

Bailey, R. 2003. "The Battle for Your Brain: Science Is Developing Ways to Boost Intelligence, Expand Memory, and More. But Will You Be Allowed to Change Your Own Mind?" *Reason,* February. http://nootropics.com/smartdrugs/nootropic.html (accessed July 8, 2008).

Baker, D. J., T. Wijshake, T. Tchkonia, et al. 2011. "Clearance of p16^{Ink4a}-positive senescent cells delays ageing-associated disorders." *Nature* 479: 232–36.

Baranzini, S. E., J. Mudge, J. C. van Velkinburgh, et al. 2010. "Genome, Epigenome, and RNA Sequences of Monozygotic Twins Discordant for Multiple Sclerosis." *Nature* 464: 1351–56.

Beauchamp, T. L., and J. F. Childress. 2001. *Principles of Biomedical Ethics.* 5th ed. New York: Oxford University Press.

Benner, S. A., D. Hutter, and A. M. Sismour. 2003. "Synthetic Biology with Artificially Expanded Genetic Information Systems: From Personalized Medicine to Extraterrestrial Life." *Nucleic Acids Research Supplement.* 3: 125–26.

Berreby, D. 2011. "Environmental Impact." *Scientist* 3: 40–44.

Biological and Environmental Research Information System. 2009. "Human Genome Project: Is Genetic Testing Regulated?" US Department of Energy Genome Programs. http://www.ornl.gov/sci/techresources/Human_Genome/medicine/genegene.shtml#regulate (accessed August 26, 2009).

Boldt, J., and O. Müller. 2008. "Newtons of the Leaves of Grass. *Nature Biotechnology* 26: 387–88.

Bonné, J. 2003. "'Go Pills': A War on Drugs?" MSNBC, January 9. http://www.msnbc.msn.com/id/3071789/ (accessed May 18, 2011).

Bova, Dr. B. 1998. *Immortality: How Science Is Extending Your Life Span—and Changing the World.* New York: Avon Books. http://www.sff.net/people/benbova/imm.html (accessed June 11, 2012).

Bradley, J. T. 2010. "Synthetic Biology: A 'Synbio-Ethics' Needed?" *Journal of the Alabama Academy of Science* 81 (3), in Annual Meeting Abstracts, Section 7. http://findarticles.com/p/articles/mi_hb178/is_2_80/ai_n35679534/ (accessed June 11, 2012).

Braun, S. 2000. *The Science of Happiness: Unlocking the Mysteries of Mood.* New York: Wiley.

Breakthrough of the Year: The Runners-Up. 2011. "Removing Old Cells to Stay Young?" *Science* 334: 1635.

Brennan, P. 2011. "First Stem-Cell Patient Reports Feeling in Legs." *Orange County Register,* April 28. http://sciencedude.ocregister.com/2011/04/28/1st-stem-cell-patient-reports-feeling-in-legs/127081/ (accessed April 26, 2012).

Bronowski, J. 1973. *The Ascent of Man.* Boston: Little, Brown and Company.

Bruder, C.E.G., A. Piotrowski, A.A.C.J. Gijsbers, et al. 2008. "Phenotypically Concordant and Discordant Monozygotic Twins Display Different DNA Copy-Number-Variation Profiles." *American Journal of Human Genetics* 82 (3): 763–71.

Bureau of Labor Statistics. 2007. "National Census of Fatal Occupational Injuries in 2006." US Department of Labor. http://www.bls.gov/news.release/archives/cfoi_08092007.pdf (accessed April 26, 2012).

Byrne, J. A., D. A. Pedersen, L. L. Clepper, M. Nelson, W. G. Sanger, S. Gokhale, D. P. Wolf, and S. M. Mitalipov. 2007. "Producing Primate Embryonic Stem Cells by Somatic Cell Nuclear Transfer." *Nature* 450: 497–502.

Campbell, A. M. 2005. "Meeting Report: Synthetic Biology Jamboree for Under-graduates." *Cell Biology Education* 4: 19–23.

Campbell, C. S. 1998. "A Comparison of Religious Views on Cloning." http://bioethics.georgetown.edu/nbac/pubs/cloning2/cc4.pdf (accessed June 10, 2012). Published in 1998 as *Cloning: Science and Society,* edited by Gary E. McDuen. Hudson, WI: Gary E. McCuen.

Caplan, A. L. 2000. "What's Morally Wrong with Eugenics?" In *Controlling Our Destinies,* edited by P. Sloan, 209–22. Notre Dame, IN: University of Notre Dame Press.

Carreau, M. 2006. "Therapy for Fanconi Anemia." In *Molecular Mechanisms of Fanconi Anemia,* edited by S. I. Ahmad and S. H. Kirk. Georgetown, TX: Landes Bioscience/Eurekah.com.

Cavalli-Sforza, L. L., P. Menozzi, and A. Piazza. 1994. *The History and Geography of Human Genes.* Princeton, NJ: Princeton University Press.

Caven, A. D. 1992. "Global Ecology Review." *Science Books and Films* 28 (2): 42.

Collins, F. 2004. "What We Do and Don't Know about 'Race,' 'Ethnicity,' Genetics, and Health at the Dawn of the Genome Era." *Nature Genetics Suppl.* 36 (11): S13–S15.

Collins, F. S., and D. Galas. 1993. "A New Five-Year Plan for the U.S. Human Genome Project." *Science* 262: 43–46.

Collins, F. S., A. Patrinos, E. Jordan, et al. 1998. "New Goals for the U.S. Human Genome Project." *Science* 282: 682–89.

Commonwealth of Australia. 2008. Prohibition of Human Cloning for Reproduction Act 2002. http://www.comlaw.gov.au/comlaw/Legislation/ActCompilation1 .nsf/0/ED63FF59CFEBB728CA2575280008CFC5?OpenDocument (accessed July 6, 2008).

Conger, K. 2011. "Scientists Turn Human Skin Cells Directly into Neurons, Skipping iPS Stage." *Stanford School of Medicine News.* http://med.stanford.edu/ism/2011 /may/wernig.html (accessed May 16, 2012).

Conrad, D. F., M. Jadobsson, G. Coop, et al. 2006. "A Worldwide Survey of Haplotype Variation and Linkage Disequilibrium in the Human Genome." *Nature Genetics* 38: 1251–60.

Crick, F. 1956. "Ideas on Protein Synthesis." National Library of Medicine's Profiles in Science. http://profiles.nlm.nih.gov/SC/B/B/F/T/_/scbbft.pdf (accessed July 23, 2009).

———. 1990. *What Mad Pursuit: A Personal View of Scientific Discovery.* New York: Basic Books.

Dees, R. H. 2008. "Better Brains, Better Selves? The Ethics of Neuroenhancements." *Kennedy Institute of Ethics Journal* 17 (4): 371–95.

Department of Health, United Kingdom. 2001. Human Reproductive Cloning Act 2001. http://www.legislation.gov.uk/ukpga/2001/23/contents (accessed April 26, 2012).

Directorate General for Internal Policies, European Parliament. 2009. "Human Enhancement Study." Policy Department A: Economic and Scientific Policy, Science and Technology Options Assessment. Presented at the workshop Human Enhancement: The Ethical Issues, for the European Parliament, Brussels, April 26, 2012.

Doherty, P. 2008. "A Mind Is a Terrible Thing to Waste." *ScienceAlert,* May 26. http:// www.sciencealert.com.au/opinions/20082705–17386–2.html (accessed July 30, 2008).

Dolgin, E. 2009. "Epigenetic Suicide Note." *Scientist* 23 (8): 18–20.

Domasio, A. 1999. *The Feeling of What Happens.* New York: Harcourt.

Donoghue, P.C.J., and J. B. Antcliffe. 2010. "Origins of Multicellularity." *Nature* 466: 41–42.

Dorff, Elliot N. 2001. "Embryonic Stem Cell Research: The Jewish Perspective (Responsum Presented to the Committee on Jewish Law and Standards, December)." *United Synagogue of Conservative Judaism Review* (Spring). http://www.oca .org/QA.asp?ID=68&SID=3 (accessed January 30, 2008).

Dreger, A. D. 2000. "Metaphors of Morality in the Human Genome Project." In *Controlling Our Destinies: Historical, Philosophical, Ethical, and Theological Perspectives on the Human Genome Project,* edited by P. R. Sloan. Notre Dame, IN: University of Notre Dame Press.

Drmanac, R. 2012. "The Ultimate Genetic Test." *Science* 336: 1110–12.

Drug Information Online. 2003. "FDA Approves Humatrope for Short Stature." http://www.drugs.com/news/fda-approves-humatrope-short-stature-3357.html (accessed April 27, 2012).

Duster, T. 2003. "The Hidden Eugenic Potential of Germ-Line Interventions." In *Designing Our Descendants—the Promises and Perils of Genetic Modifications,* edited by A. R. Chapman and M. S. Frankel, 156–78. Baltimore, MD: Johns Hopkins University Press.

Dyson, F. 2007. "Our Biotech Future." *New York Review of Books* 54 (12), July 19. http://www.nybooks.com/articles/archives/2007/jul/19/our-biotech-future/?pagination=false&printpage=true (accessed June 15, 2012).

El Albani, A., S. Bengstson, D. E. Canfield, et al. 2010. "Large Colonial Organisms with Coordinated Growth in Oxygenated Environments 2.1 Gyr Ago." *Nature* 466: 100–104.

Emonson, D. L., and R. D. Vanderbeek. 1995. "The Use of Amphetamines in U.S. Air Force Tactical Operation during Desert Shield and Storm." *Aviation, Space, and Environmental Medicine* 66 (8): 802.

Esteller, M. 2011. "Epigenetic Changes in Cancer." *Scientist* 3: 34–39.

ETC Group. 2007. "Extreme Genetic Engineering: An Introduction to Synthetic Biology." www.etcgroup.org (accessed February 3, 2009).

———. 2010. "New Directions: The Ethics of Synthetic Biology and Emerging Technologies." Letter to Presidential Commission for the Study of Bioethical Issues. http://www.etcgroup.org/upload/publication/pdf_file/Civil%20Society%20Letter%20to%20Presidents%20Commission%20on%20Synthetic%20Biology_0.pdf (accessed March 23, 2011).

Ferber, D. 2004. "Microbes Made to Order." *Science* 303: 158–61.

Federal Emergency Management Agency. 2010. National Environmental Policy Act (NEPA). US Department of Homeland Security. http://www.fema.gov/plan/ehp/ehplaws/nepa.shtm#2 (accessed March 26, 2011).

Foekens, J. A., Y. Wang, J. W. Martens, et al. 2008. "The Use of Genomic Tools for the Molecular Understanding of Breast Cancer and to Guide Personalized Medicine." *Drug Discovery Today* 13 (11–12): 481–87.

Ford, N. M. 1988. *When Did I Begin? Conception of the Human Individual in History, Philosophy and Science.* Cambridge, UK: Cambridge University Press.

Franjfurt, H. 2006. "Taking Ourselves Seriously." In *Taking Ourselves Seriously and Getting It Right,* edited by D. Satz, 1–26. Stanford: Stanford University Press.

Fraser, C. M., M. D. Adams, J. D. Gocayne, et al. 2003. US Patent Application 20030170663. US Patent and Trademark Office. http://appft.uspto.gov/netacgi

/nph-Parser?Sect1=PTO1&Sect2=HITOFF&d=PG01&p=1&u=%2Fnetahtml
%2FPTO%2Fsrchnum.html&r=1&f=G&l=50&s1=%2220030170663%22.PGNR
.&OS=DN/20030170663&RS=DN/20030170663 (accessed April 26, 2012).

Frye, I. 2000. *The Emergence of Life on Earth: A Historical and Scientific Overview.* Brunswick, NJ: Rutgers University Press.

Fukuyama, F. 2002. *Our Posthuman Future—Consequences of the Biotechnology Revolution.* 1st ed. New York: Farrar, Straus and Giroux.

Gallup. 2003. "Detailed Survey Results." Center for Genetics and Society. http://www.geneticsandsociety.org/article.php?id=404#2003gallup (accessed June 22, 2009).

Galton, Francis. 1883. *Inquiries into Human Faculty and Its Development.* London: Macmillan.

Garreau, J. 2005. *Radical Evolution: The Promise and Peril of Enhancing Our Minds, Our Bodies—and What It Means to Be Human.* New York: Doubleday.

Gazzaniga, M. S. 2005. "The Thoughtful Distinction between Embryo and Human." *Chronicle of Higher Education,* April 8, B10–B12.

Gibson, D. G., G. A. Benders, K. C. Axelrod, et al. 2008. "One-Step Assembly in Yeast of 25 Overlapping DNA Fragments to Form a Complete Synthetic Mycoplasm Genitalium Genome." *Proceedings of the National Academy of Sciences USA* 105 (51): 20404–9.

Gibson, D. G., et al. 2010. "Creation of a Bacterial Cell Controlled by a Chemically Synthesized Genome." *Science* 329 (5987): 52–56.

Gilbert, S. F., A. L. Tyler, and E. J. Zackin. 2005. "Genetic Essentialism." In *Bioethics and the New Embryology: Springboards for Debate,* 227–40. Sunderland, MA: Sinauer Associates.

Glass, J. I., N. Assad-Garcia, N. Alperovich, et al. 2006. "Essential Genes of a Minimal Bacterium." *Proceedings of the National Academy of Sciences USA* 103: 425–30.

Goljan, E. F. 2004. *Pathology, Rapid Review Series.* Philadelphia, PA: Mosby Books, Elsevier, Health Sciences Division.

Great Books film series. 1993. *The Origin of Species: Beyond Genesis.* Silver Springs, MD: Discovery Communications.

Green, R. E., J. Krause, A. W. Briggs, et al. 2010. "A Draft Sequence of the Neandertal Genome." *Science* 328: 710–22.

Greiss, S., and J. W. Chin. 2011. "Expanding the Genetic Code of an Animal." *Journal of the American Chemical Society* 133: 14196–99.

Gurdon, J. B., and V. Uehlinger. 1966. "'Fertile' Intestine Nuclei." *Nature* 210: 1240–41.

Haidt, J. 2006. *The Happiness Hypothesis: Finding Modern Truth in Ancient Wisdom.* New York: Basic Books.

Hardesty, G. 2008. "The Baby Born to Save Her Sister Says She Has No Regrets." *Orange County Register,* July 21. http://www.ocregister.com/articles/marissa-anissa-ayala-2100465-marrow-story# (accessed July 4, 2009).

Harris, J. 1985. *The Value of Life: An Introduction to Medical Ethics.* New York: Routledge.

———. 1998. *Clones, Genes, and Immortality: Ethics and the Genetic Revolution.* Oxford: Oxford University Press.

Hayflick, L., and P. S. Moorhead. 1961. "The Serial Cultivation of Human Diploid Cell Strains." *Experimental Cell Research* 25 (3): 585–621.

Hazen, R. 2005. *Genesis: The Scientific Quest for Life's Origins.* Washington, DC: National Academies Press.

Hippocratic Oath. 2012. http://www.pbs.org/wgbh/nova/body/hippocratic-oath-today.html (accessed May 7, 2012).

Hirst, K. K. 2009. "Human Migration Map." http://archaeology.about.com/od/stoneage/ss/tishkoff_2.htm (accessed September 6, 2009).

Hollmer, M. 2009. "Brain-Computer Interface, Developed at Brown, Begins New Clinical Trial." Brown University, June 10. http://news.brown.edu/pressreleases/2009/06/braingate2?printable (accessed September 2, 2009).

Hooper, J., and D. Teresi. 1987. *The Three-Pound Universe.* New York: McMillan.

Howard Hughes Medical Institute. 2011. "HHMI Investigators: Jack W. Szostak, Ph.D." http://www.hhmi.org/research/investigators/szostak_bio.html (accessed October 3, 2011).

Huxley, A. 1932. *Brave New World.* New York: HarperCollins.

Iles, G. 2003. *The Footprints of God.* New York: Simon & Schuster, Inc.

International Human Genome Sequencing Consortium. 2001. "Initial Sequencing and Analysis of the Human Genome." *Nature* 409: 860–921.

———. 2004. "Finishing the Euchromatic Sequence of the Human Genome." *Nature* 431: 931–45.

Jastrow, R., and M. Rampino. 2008. *Origins of Life in the Universe.* Cambridge: Cambridge University Press.

Jones, P. A., and R. Martienssen. 2005. "A Blueprint for a Human Epigenome Project: The AACR Human Epigenome Workshop." *Cancer Research* 65 (24): 11241–46.

Jones, Rev. David. 2001. "A Theologian's Brief on the Place of the Human Embryo within the Christian Tradition." *Ethics & Medicine: An International Journal of Bioethics* 17: 3.

Journal of Gene Medicine. "Gene Therapy Clinical Trials Worldwide." http://www.abedia.com/wiley/index.html (accessed April 26, 2012).

Journal Sentinel PolitiFact. "Scott Walker Says Scientist Agree That Adult Stem Cell

Research Holds Greater Promise Than Embryonic Stem Cell Research." http://www.politifact.com/wisconsin/statements/2010/oct/21/scott-walker/scott-walker-says-scientists-agree-adult-stem-cell/ (accessed June 10, 2012).

Kaiser, J. 2007. "Attempt to Patent Artificial Organism Draws a Protest." *Science* 316: 1557.

Kaji, K., K. Norrby, A. Paca, et al. 2009. "Virus-Free Induction of Pluripotency and Subsequent Excision of Reprogramming Factors." *Nature* 458: 771–75.

Kant, Immanuel. [1785] 2000. In *The Moral Life: An Introductory Reader in Ethics and Literature,* edited by Louis P. Pojman and Lewis Vaughn, 297. New York: Oxford University Press.

Karp, G. 2010. "Diseases that Result from Abnormal Mitochondrial or Peroxisomal Function." In *Cell and Molecular Biology: Concepts and Experiments.* 6th ed., 201–3. Hoboken, NJ: Wiley & Sons.

Kass, L. 1997. "The Wisdom of Repugnance." *New Republic* 216 (22): 17–26.

Katsnelson, A. 2009. "Brick by Brick." *Scientist* 23 (2): 42–47. http://classic.the-scientist.com/article/display/55378/ (accessed June 13, 2012).

Kendall, S. K. 2009. "Cloning—A Webliography, Laws and Public Policy about Cloning." http://staff.lib.msu.edu/skendall/cloning/laws.htm (accessed July 6, 2009).

Kim, J. B., B. Gerber, M. J. Araúzo-Bravo, et al. 2009. "Direct Reprogramming of Human Neural Stem Cells by OCT4." *Nature* 461: 649–53.

Kitzman, J. O., M. W. Snyder, M. Ventura, et al. 2012. "Noninvasive Whole-Genome Sequencing of a Human Fetus." *Science Translational Medicine* 4: 137ra76.

Klipstein, S. 2005. "Preimplantation Genetic Diagnosis: Technological Promise and Ethical Perils." *Fertility and Sterility* 83: 1347–1534.

Klotz, I. 2009. "Synthetic Life Form Grows in Florida Lab. Creator: First Synthetic Genetic System Capable of Darwinian Evolution." MSNBC Discovery News, February 27. http://www.msnbc.msn.com/id/29430688/ (accessed April 26, 2012).

Kolata, G. 1997. "With Cloning of a Sheep, the Ethical Ground Shifts." *New York Times,* February 27.

Kramer, P. D. 1997. *Listening to Prozac.* New York: Penguin Books.

Kriebel, D., J. Tickner, P. Epstein, et al. 2001. "The Precautionary Principle in Environmental Science." *Environmental Health Perspectives* 109 (9): 871–76. http://ehp03.niehs.nih.gov/article/fetchArticle.action;jsessionid=CCAF7561D8939F38BB7979E5564D33BD?articleURI=info%3Adoi%2F10.1289%2Fehp.01109871#r3 (accessed March 26, 2011).

Krieger, N., J. T. Chen, P. D. Waterman, et al. 2005. "Painting a Truer Picture of US Socioeconomic and Racial/Ethnic health Inequalities: The Public Health Dispari-

ties Decoding Project." *American Journal of Public Health* 95: 312–23. http://www
.ajph.org/cgi/reprint/95/2/312 (accessed September 5, 2009).

Kubicek, S. 2011. "Epigenetics: A Primer." *Scientist* 3: 32–33.

Lalande, M. 1996. "Parental Imprinting and Human Disease." *Annual Review of Genetics* 30: 173–95.

Lamason, R. L., M.P.K. Mohideen, J. R. Mest, et al. 2005. "SLC24A5, a Putative Cation Exchanger, Affects Pigmentation in Zebrafish and Humans." *Science* 310: 1782–86.

Lamont, J., and C. Favor. 2009. "Distributive Justice." In *Stanford Encyclopedia of Philosophy,* edited by E. N. Zalta. Metaphysics Research Lab. http://plato.stanford.edu/entries/justice-distributive/ (accessed April 26, 2012).

Lartigue, C., J. I. Glass, N. Alperovich, et al. 2007. "Genome Transplantation in Bacteria: Changing One Species to Another." *Science* 317: 632–37.

Lartigue, C., S. Vasshee, M. A. Alfire, et al. 2009. "Creating Bacterial Strains from Genomes That Have Been Cloned and Engineered in Yeast." *Science* 325 (5948): 1693–96.

Leakey, R., and R. Lewin. 1995. *The Sixth Extinction: Patterns of Life and the Future of Humankind.* New York: Random House.

LeDoux, J. 2002. *Synaptic Self: How Our Brains Become Who We Are.* New York: Penguin Books.

Leopold, Aldo. 1949. "The Land Ethic." In *A Sand County Almanac,* 1–92. New York: Oxford University Press.

Lehrer, J. 2009. "Neuroscience: Small, Furry . . . and Smart." *Nature* 461: 862–64.

Levy, S., G. Sutton, P. C. Nq, et al. 2007. "The Diploid Genome Sequence of an Individual Human." *PLoS Biology* 5 (10): e254. http://www.plosbiology.org/article/info:doi/10.1371/journal.pbio.0050254 (accessed April 26, 2012).

Lim, K. S., P. Kwan, and C. T. Tan. 2008. "Association of HLA-B*1502 Allele and Carbamazepine-Induced Severe Adverse Cutaneous Drug Reaction among Asians, a Review." *Neurology Asia* 13: 15–21.

Lin, P., K. Abney, and G. A. Bekey. 2012. *Robot Ethics: The Ethical and Social Implications of Robotics.* Cambridge, MA: MIT Press.

Lovejoy, A. O. 1936. *The Great Chain of Being: A Study of the History of an Idea.* Cambridge: Harvard University Press.

Martuzzi, M., and J. A. Tickner. 2004. *The Precautionary Principle: Protecting Public Health, the Environment and the Future of Our Children.* Coopenhagen, Denmark: World Health Organization. http://www.euro.who.int/__data/assets/pdf_file/0003/91173/E83079.pdf (accessed June 14, 2012).

Mayr, E. 1997. *This Is Biology: The Science of the Living World.* Cambridge, MA: Belknap Press.

McGowan, P. O., A. Sasak, A. C. D'Alessio, et al. 2009. "Epigenetic Regulation of the Glucocorticoid Receptor in Human Brain Associates with Childhood Abuse." *Nature Neuroscience* 12: 342–48.

McKenny, G. P. 2002. "Religion and Gene Therapy: The End of One Debate, the Beginning of Another." In *A Companion to Genethics,* edited by J. Burley and J. Harris, 287–301. Molden, MA: Blackwell Publishers.

Morrow, K. J. 2009. "Novel Systems Zero in on the $100 Genome." *Genetic Engineering & Biotechnology News* 29 (8). April 15 feature article. http://www.genengnews .com/gen-articles/novel-systems-zero-in-on-the-100-genome/2864/ (accessed June 20, 2012).

Murray, T. H. 2007. "Enhancement." In *The Oxford Handbook of Bioethics,* edited by B. Steinbock, 491–515. New York: Oxford University Press.

Murthy, V. H., H. M. Krumholz, and C. P. Gross. 2004. "Participation in Cancer Clinical Trials: Race-, Sex-, and Age-Based Disparities." *Journal of the American Medical Association* 291: 2720–26.

National Bioethics Advisory Commission. 1997. "Cloning Human Beings." Letter from the president and chap. 3, "Religious Perspectives." http://bioethics .georgetown.edu/nbac/pubs/cloning1/cloning.pdf (accessed June 10, 2012).

National Conference of State Legislatures. 2008a. "Embryonic and Fetal Research Laws: Stem Cell Research." http://www.ncsl.org/default.aspx?tabid=14413 (accessed November 11, 2011).

———. 2008b. "Human Cloning Laws." http://www.ncsl.org/IssuesResearch/Health /HumanCloningLaws/tabid/14284/Default.aspx (accessed July 6, 2009).

National Council of State Legislatures. 2008. "Human Cloning Laws." Updated January 2008. http://www.ncsl.org/IssuesResearch/Health/HumanCloningLaws /tabid/14284/Default.aspx (accessed July 6, 2009).

National Institutes of Health. 1994. "Report of the Human Embryo Research Panel." http://bioethics.georgetown.edu/pcbe/reports/past_commissions/human_embryo _vol_1.pdf (accessed January 16, 2012).

———. 2007. "Overview of the Policy and Biosafety Framework for Human Gene Transfer Research: The NIH Guidelines for Research Involving Recombinant DNA Molecules." http://oba.od.nih.gov/oba/IBC/ASGT_2007_Training/Overview %20of%20the%20NIH%20Guidelines.pdf (accessed March 26, 2011).

———. 2009. "National Institutes of Health Guidelines on Human Stem Cell Research." NIH Stem Cell Information. http://stemcells.nih.gov/policy/2009guidelines .htm (accessed October 12, 2009).

————. 2011. "About the International HapMap Project." National Human Genome Research Institute. http://www.genome.gov/11511175#al-4 (accessed May 18, 2011).

————. 2012. "NIH Guidelines for Research Involving Recombinant DNA Molecules." http://oba.od.nih.gov/rdna/nih_guidelines_oba.html (accessed May 16, 2012).

Nemade, R. 2005. "Human 'Epigenome' Project." HUM-MOLGEN, December 21. http://hum-molgen.org/NewsGen/12–2005/000025.html (accessed August 24, 2009).

Newport, F. 2009. "On Darwin's Birthday, Only 4 in 10 Believe in Evolution: Belief Drops to 24% among Frequent Church Attenders." Gallup. http://www.gallup.com/poll/114544/darwin-birthday-believe-evolution.aspx (accessed March 26, 2011).

News-Medical.Net. 2009. "Geron Plans to Advance Clinical Program for Spinal Cord Injury." October 31. http://www.news-medical.net/news/20091031/Geron-plans-to-advance-clinical-program-for-spinal-cord-injury.aspx (accessed July 23, 2010).

Nobel, M. 2010. "Stem Cells: Their Potential for Treating PD." Parkinson's Disease Foundation. http://www.pdf.org/en/spring05_Stem_Cells (accessed May 18, 2011).

Noonan, J. P., G. Coop, S. Kudaravalli, et al. 2006. "Sequencing and Analysis of Neanderthal Genomic DNA." *Science* 314 (5802): 1113–18.

Nordenfelt, L. 1995. *On the Nature of Health: An Action-Theory Approach.* 2nd ed. Dordrecht, Netherlands: Kluwer Academic Publishers.

O'Connor, A. 2008. "The Claim: Identical Twins Have Identical DNA." *New York Times,* March 11.

Olshansky, S. J., and B. A. Carnes. 2001. *The Quest for Immortality.* New York: W. W. Norton & Company.

Open Source Initiative. 2006. "The Open Source Definition (Annotated)." California Public Benefit Corporation. http://www.opensource.org/osd.html (accessed April 20, 2009).

Paddock, C. 2011. "Female Smokers at Higher Risk of Heart Disease." *Medical News Today,* August 11.

Palpant, N. J., and C. E. Murry. 2012. "Reprogramming the Injured Heart." *Nature* 485: 585–86.

Parens, E. 1998. "Is Better Always Good? The Enhancement Project." In *Enhancing Human Traits: Ethical and Social Implications,* edited by E. Parens, 1–28. Washington, DC: Georgetown University Press.

Parens, E., J. Johnston, and J. Moses. 2008. "Do We Need 'Synthetic Bioethics'?" *Science* 321: 1449.

————. 2009. "Ethical Issues in Synthetic Biology: An Overview of the Debates." Woodrow Wilson International Center for Scholars. http://www.synbioproject .org/process/assets/files/6334/synbio3.pdf (accessed June 20, 2012).

PBS Home Video. 2008. *Power of the Past with Bill Moyers: Florence.* David Gruben, director.

Pearce, D. 2009. "Soma in Aldous Huxley's Brave New World." BLTC Research. http://www.huxley.net/soma/somaquote.html (accessed September 1, 2009).

Peeper, D. S. 2011. "Ageing: Old Cells under Attack." *Nature* 479: 186–87.

Pennisi, E. 2003. "Venter Cooks Up a Synthetic Genome in Record Time." *Science* 302 (5649): 1307.

————. 2009. "Two Steps Forward for Synthetic Biology." *Science* 325: 928–29.

The People of the United Methodist Church. 2001. "Ethics of Embryonic Stem Cell Research" (a resolution). General Conference of the United Methodist Church. http://archives.umc.org/interior.asp?ptid=4&mid=6560 (accessed June 1, 2012).

Pera, M. F. 2009. "Low-Risk Reprogramming." *Nature* 458: 715–16.

————. 2011. "Stem Cells: The Dark Side of Induced Pluripotency." *Nature* 471: 46–47.

Peters, T., and G. Bennett. 2003. "A Plea for Beneficence: Reframing the Embryo Debate." In *God and the Embryo,* edited by B. Waters and R. Cole-Turner. Washington, DC: Georgetown University Press.

Pico della Mirandola, G. [1486] 2005. "Oration on the Dignity of Man." In *The Human Odyssey—Readings from Original Sources,* edited by J. T. Bradley and Human Odyssey Faculty, translated by E. L. Forbes, 286–87. Boston, MA: Pearson Custom Publishing.

Plato. 1996. "Meno." In *Plato: The Collected Dialogues,* edited by E. Hamilton and H. Gairns, translated by W.K.C. Guthrie, 353–84. Princeton, NJ: Princeton University Press.

Pollack, A. 2009. "F.D.A. Approves a Stem Cell Trial." *New York Times,* January 23.

Pontifical Academy for Life, Vatican City. 2000. "Declaration on the Production and the Scientific and Therapeutic Use of Human and Embryonic Stem Cells." http:// www.vatican.va/roman_curia/pontifical_academies/acdlife/documents/rc_pa _acdlife_doc_20000824_cellule-staminali_en.html (accessed January 30, 2008).

Presbyterian Church (USA). 2001. Actions of the 213th General Assembly from the Office of the General Assembly. "Attachment A: Statement on the Ethical and Moral Implications of Stem Cell and Fetal Tissue Research." http://www.pcusa .org/oga/actions-of-213.htm#attachment (accessed January 30, 2008).

Presidential Commission for the Study of Bioethical Issues. 2010. "New Directions: The Ethics of Synthetic Biology and Emerging Technologies, Washington, D.C." http://www.bioethics.gov/documents/synthetic-biology/PCSBI-Synthetic -Biology-Report-12.16.10.pdf (accessed April 27, 2012).

President's Council on Bioethics. 2003a. "Age-Retardation: Scientific Possibilities and Moral Challenges." http://www.bioethics.gov/background/age_retardation .html (accessed September 16, 2009).

———. 2003b. *Beyond Therapy: Biotechnology and the Pursuit of Happiness.* New York: Harper Collins.

———. 2005. "White Paper: Alternative Sources of Pluripotent Stem Cells." Washington, DC. http://bioethics.georgetown.edu/pcbe/reports/white_paper/index .html (accessed July 24, 2010).

Rauscher, R. J., III. 2005. "It Is Time for a Human Epigenome Project." *Cancer Research* 65 (24): 11229. http://cancerres.aacrjournals.org/content/65/24/11229 .full (accessed April 23, 2011).

Rawls, John. 1993. *Political Liberalism.* New York: Columbia University Press.

Ro, D-K., E. M. Paradise, M. Ouellet, et al. 2006. "Production of the Antimalarial Drug Precursor Artemisinic Acid in Engineered Yeast." *Nature* 440: 940–43.

Roberts, J. C. 2002. "Customizing Conception: A Survey of Preimplantation Genetic Diagnosis and the Resulting Social, Ethical, and Legal Dilemmas." *Duke Law and Technology Review.* http://www.law.duke.edu/journals/dltr/articles/2202dltr0012 .html (accessed June 7, 2012).

Robertson, J. A. 1994. *Children of Choice: Freedom and the New Reproductive Technologies.* Princeton, NJ: Princeton University Press.

Rudman, D., A. G. Feller, H. S. Nagraj, et al. 1990. "Effects of Human Growth Hormone in Men over 60 Years Old." *New England Journal of Medicine* 323: 1–6.

Sandel, M. J. 2007. *The Case against Perfection: Ethics in the Age of Genetic Engineering.* Cambridge, MA: Belnap Press.

Sankar, P., and J. Kahn. 2005. "BiDil: Race Medicine or Race Marketing?" Health Affairs—Web Exclusive, Project HOPE—People-to-People Health Foundation. http://content.healthaffairs.org/cgi/reprint/hlthaff.w5.455v1.pdf (accessed September 10, 2009).

Savulescu, J. 2007. "Genetic Interventions and the Ethics of Enhancement of Human Beings." In *The Oxford Handbook of Bioethics,* edited by B. Steinbock, 516–35. New York: Oxford University Press.

Schmidt, M., H. Torgersen, A. Ganquili-Mitra, et al. 2008. "SYNBIOSAFE e-Conference: Online Community Discussion on the Societal Aspects of Synthetic Biology." *Systems and Synthetic Biology* 2: 7–17. http://www.springerlink.com/content /rtt71124tn6096kh (accessed April 27, 2012).

Schmitz, R. J., M. D. Schultz, M. G. Lewsey, et al. 2011. "Transgenerational Epigenetic Instability Is a Source of Novel Methylation Variants." *Science* 334 (6054): 369–73.

Science Daily. 2011. "Are Genes Our Destiny? Scientists Discover 'Hidden' Code in DNA Evolves More Rapidly Than Genetic Code." September 16. http://www.sciencedaily.com/releases/2011/09/110916152401.htm (accessed September 21, 2011).

Science New Focus. "Mysteries of the Cell." *Science* 334 (2011): 1046–51.

Scientific American. 1991. "When Did Eukaryotic Cells (Cells with Nuclei and Other Internal Organelles) First Evolve? What Do We Know about How They Evolved from Earlier Life-Forms?" http://www.scientificamerican.com/article.cfm?id=when-did-eukaryotic-cells&page=2 (accessed July 15, 2010).

Scudellari, M. 2012. "Same Day Genomes." *Scientist,* January 13. http://the-scientist.com/2012/01/13/same-day-genomes/ (accessed January 17, 2012).

Sharp, M. 2011. "No More Science Fiction—HIV Gene Therapy Delivers." HIV and Hepatitis.com Coverage of the 18th Conference on Retroviruses and Opportunistic Infections (CROI 2011). http://www.hivandhepatitis.com/2011_conference/croi2011/docs/0304_2010a.html (accessed January 20, 2012).

Shubin, N. 2009. *Your Inner Fish: A Journey into the 3.5-Billion Year History of the Human Body.* New York: Vintage Books.

Skloot, R. 2010. *The Immortal Life of Henrietta Lacks.* New York: Random House.

Sloan, Phillip R., ed. 2000. *Historical, Philosophical, Ethical, and Theological Perspectives on the Human Genome Project.* Notre Dame, IN: University of Notre Dame Press.

Smith, H. O., C. A. Hutchison III, C. Pfannkoch, and J. C. Venter. 2003. "Generating a Synthetic Genome by Whole Genome Assembly: φX174 Bacteriophage from Synthetic Oligonucleotides." *Proceeding of the National Academy of Sciences USA* 100: 15440–45.

South and Meso American Indian Rights Center. 1995. "Human Genome Diversity Project: Declaration of Indigenous Peoples of the Western Hemisphere Regarding the Human Genome Diversity Project." http://www.indians.org/welker/genome/htm (accessed September 10, 2009).

Southern Baptist Convention. 1999. Resolution #17: Human Embryonic and Stem Cell Research, adopted at the SBC on June 16. www.johnstonsarchive.net/baptist/sbcares.html (accessed January 30, 2008).

Steele, M., and D. Penny. 2010. "Common Ancestry Put to the Test." *Nature* 465: 168–69.

Symmans, W. F., C. Hatzis, C. Sotiriou, et al. 2010. "Genomic Index of Sensitivity to Endocrine Therapy for Breast Cancer." *Journal of Clinical Oncology* 28 (27): 4101–3.

Taaffe, D. R., L. Pruitt, J. Reim, et al. 1994. "Effect of Recombinant Human Growth Hormone on the Muscle Strength Response to Resistance Exercise in Elderly Men." *Journal of Clinical Endocrinology and Metabolism* 79: 1361–66.

Tate, S. K., and D. B. Goldstein. 2004. "Will Tomorrow's Medicines Work for Everyone?" *Nature Genetics Suppl.* 36 (11): S34–S42.

Than, K. 2007. "Breakthrough Could Lead to Artificial Life Forms." *LiveScience,* June 28. http://www.livescience.com/7303-breakthrough-lead-artificial-life-forms.html (accessed April 27, 2012).

Theobald, D. L. 2010. "A Formal Test of the Theory of Universal Common Ancestry." *Nature* 465: 219–22.

Tooley, M. 1983. *Abortion and Infanticide.* New York: Oxford University Press.

Tully, T., R. Bourtchouladze, R. Scott, and J. Tallman. 2003. "Targeting the CREB Pathway for Memory Enhancers." *Nature Reviews Drug Discovery* 2: 267–77.

Turner, L. 2004. "Is Repugnance Wise? Visceral Responses to Biotechnology." *Nature Biotechnology* 22: 269–70.

University of Minnesota. 2001. "Umbilical Cord Transplant Succeeds for Molly Nash." UM News Release, January 4. http://www1.umn.edu/news/news-releases/2001/UR_RELEASE_MIG_1251.html (accessed January 11, 2011).

US National Academy of Sciences. 2010. "Books and Reports." Evolution Resources. http://www.nationalacademies.org/evolution/Reports.html (accessed March 26, 2011).

US Social Security Administration. 2011. "Period Life Table." Social Security Online. Actuarial Publications. http://www.ssa.gov/OACT/STATS/table4c6.html (accessed June 20, 2012).

Veblen, T. 1914. *The Instinct of Workmanship and the State of the Arts.* New York: MacMillan Company.

Venter, J. C., M. D. Adams, E. W. Myers, et al. 2001. "The Sequence of the Human Genome." *Science* 291: 1304–51.

Vogel, G. 1999. "Breakthrough of the Year: Capturing the Promise of Youth." *Science* 286 (5448): 2238.

Wadman, M. 2008. "James Watson's Genome Sequenced at High Speed." *Nature* 452: 788.

———. 2011. "Court Quashes Stem-Cell Lawsuit: US Judge Throws Out Case Meant to Halt Federal Funding, but Research Remains Vulnerable." *Nature* 476: 14–15.

Walters, L., and J. G. Palmer. 1997. *The Ethics of Human Gene Therapy.* New York: Oxford University Press.

Westoff, C. F., and R. R. Rindfuss. 1974. "Sex Preselection in the United States: Some Implications." *Science* 184: 633–36.

Wilmut, I., A. E. Schnieke, J. McWhir, A. J. Kind, and K. H. Campbell. 1997. "Viable Offspring Derived from Fetal and Adult Mammalian Cells." *Nature* 385 (6619): 810–13.

Wilson, E. O. 2006. *The Creation: An Appeal to Save Life on Earth.* New York: W. W. Norton & Company.

Wolf, P. D. 2003. "Apparatus for Acquiring and Transmitting Neural Signals and Related Methods." US Patent 7187968. http://www.freepatentsonline.com/7187968 .html (accessed September 2, 2009).

Wolpe, P. R. 2002. "Treatment, Enhancement, and the Ethics of Neurotherapeutics." *Brain and Cognition* 50: 393–95.

Woltjen, K., I. P. Michael, P. Mohseni, et al. 2009. "PiggyBac Transposition Reprograms Fibroblasts to Induced Pluripotent Stem Cells." *Nature* 458: 766–70.

Woodring, Tech. Sgt. J. C. 2004. "Air Force Scientists Battle Aviator Fatigue." *Air Force Print News,* US Air Force. http://www.af.mil/news/story.asp?id=123007615 (accessed May 16, 2012).

World Medical Association (WMA). 2006. "World Medical Association International Code of Medical Ethics." http://www.wma.net/en/30publications/10policies/c8/ (accessed June 7, 2012).

Yang, Z., D. Hutter, P. Sheng, A. M. Sismour, and S. A. Benner. 2006. "Artificially Expanded Genetic Information System: A New Base Pair with an Alternative Hydrogen Bonding Pattern." *Nucleic Acids Research* 34: 6095–101.

Zhu, T. G., and J. W. Szostak. 2009. "Coupled Growth and Division of Model Protocell Membranes." *Journal of the American Chemical Society* 131: 5705–13. http://pubs.acs.org/doi/full/10.1021/ja900919c (accessed October 3, 2011).

Zielinska, E. 2012. "Manipulating Eggs to Avoid Disease: A United Kingdom Ethics Council Approves Altering Human Egg Cells, Which Could Allow Doctors to Correct Mitochondrial Disease in IVF Patients." *The Scientist,* June 13. *http://the-scientist.com/2012/06/13/manipulating-eggs-to-avoid-disease/* (accessed June 15, 2012).

Websites

http://calerie.dcri.duke.edu/index.html (accessed June 20, 2012).

http://genome.ucsc.edu/cgi-bin/hgGateway (accessed June 20, 2012).

http://partsregistry.org/Main_Page (accessed June 20, 2012).

http://protocells.lanl.gov (accessed June 20, 2012).

http://www.bionanogenomics.com/ (accessed June 20, 2012).

http://www.completegenomics.com (accessed June 20, 2012).

http://www.ebi.ac.uk/genomes/eukaryota.html (accessed June 20, 2012).

http://www.epigenome.org (accessed June 20, 2012).

http://www.hapmap.org/thehapmap.html.en (accessed June 20, 2012).

http://www.ivfbabies.com/ (accessed June 20, 2012).

http://www.ncbi.nlm.nih.gov/projects/genome/guide/human/ (accessed June 20, 2012).

http://www.oxfordancestors.com (accessed June 20, 2012).

http://www.personalgenomics.org (accessed June 20, 2012).

http://www.protocell.org (accessed June 20, 2012).

Index

261, 264–65; description of, *255, 256;* ethical issues for, 269–81; future applications of, 268–69; future practitioners of, 265–66; and genetic engineering, 257–59; presidential commission report on, 275; products of, 266–68; species transformed by, 261, *262,* 263; standard biological part for, 259–60, 266; and synthetic genome, 263–64; top-down, 259, *260,* 261, *262,* 263–64. *See also* minimal genome

Tay-Sachs disease, 86
telomere, 188–89, 212, 222
TEM. *See* microscopy, transmission electron
test tube baby, 89, 182, 242
therapeutic cloning. *See* cloning, therapeutic
Thomson, James, 37, 201
totipotent cell, 56, 184, 204n6
transcription, 6, *19,* 20–21. *See also* central dogma of biology
transcription factor(s), 18, *48,* 250n9, 295
translation, 16, *19. See also* central dogma of biology
transmission electron microscopy. *See* microscopy, transmission electron

triplet codon. *See* codon
trophoectoderm, *40,* 54, 55
Turner syndrome, 87
twins: and cloning, 192, 195, 204n4; fraternal, 55; identical, 55–56, 75n6, 122, 127, 132n18; and twinning process during embryogenesis, 55

United Church of Christ, 64
United Methodist Church, 60–61

vectors for genetic engineering, 74n4, 162, *163,* 164, 178n3, 295
Venter, Craig: and Human Genome Project, 114–17; and synthetic biology, 261, *262,* 263–64, 268, 272, 284n9, 284n10, 285n19

Walker, Scott, 202
Walters, LeRoy, 167, 251n17
Watson, James, 112–13, 117
Wilmut, Ian, 184–86. *See also* Dolly
World Medical Association's International Code of Medical Ethics, 95, 108n15

Zygote, xv, 52–54, 55, 56, 183